THE COMMONWEALTH AND INTERNATIONAL LIBRARY
Joint Chairmen of the Honorary Editorial Advisory Board
SIR ROBERT ROBINSON, O.M., F.R.S., LONDON
DEAN ATHELSTAN SPILHAUS, MINNESOTA

METEOROLOGY DIVISION
General Editor: R. S. SCORER

AIR POLLUTION

FRONTISPIECE. The atmosphere seen at sunset from an aircraft making a turn at the tropopause. The troposphere below is stirred up and coloured with natural and man-made pollution: this is the region of clouds and weather. Above is the stratosphere which is clean and dry, and therefore dark blue because only a small amount of blue light is scattered by it.

Throughout the troposphere there are often other inversions through which pollution does not penetrate. Pollution below the tropopause is usually diluted quickly and removed by rain after only a few days. Pollution which penetrates into the stratosphere may remain there for years

AIR POLLUTION

BY

R. S. SCORER

Professor of Theoretical Mechanics at
Imperial College, London

PERGAMON PRESS

OXFORD · NEW YORK

TORONTO · SYDNEY · BRAUNSCHWEIG

PERGAMON PRESS LTD.,
Headington Hill Hall, Oxford

PERGAMON PRESS INC.,
Maxwell House, Fairview Park,
Elmsford, New York 10523

PERGAMON OF CANADA LTD.,
207 Queen's Quay West, Toronto 1

PERGAMON PRESS (AUST.) PTY. LTD.,
19a Boundary Street, Rushcutters Bay,
N.S.W. 2011, Australia

VIEWEG & SOHN GmbH,
Burgplatz 1, Braunschweig

First edition 1968

Reprinted 1972

Library of Congress Catalog Card No. 66-29604

Printed in Great Britain by A. Wheaton & Co., Exeter

08 012275 2 (flexicover)
08 013345 2 (hard cover)

CONTENTS

LIST OF PLATES vii

PREFACE xi

ACKNOWLEDGEMENTS xiii

1 *Pollution over Flat Country* 1

2 *Dilution: Formulae and Mechanisms* 20

3 *High Level Inversions* 54

4 *Ground and Valley Inversions* 70

5 *Wet and Coloured Plumes and Natural Pollution* 86

6 *Separation* 107

7 *Some Effects of Air Pollution* 124

8 *Repercussions* 138

INDEX 149

LIST OF PLATES

FRONTISPIECE   The atmosphere at sunset

1.1.	Shallow fog seen from the air	4
1.2.	Radiation fog	5
1.3.	Cottage smoke under an inversion	5
1.4.	Long smoke trail at dawn	6
1.5.	Smoke layer at night	6
1.6.	Sea fog dispersing inland	8
1.7.	Sea fog off Cape Cod	8
1.8.	Guttation on grass	9
1.9.	Sea fog becoming cumulus inland	9
1.10.	Mirage over concrete surface	10
1.11.	Cumulus over forest fire	11
1.12.	Cumulus formed over power station	11
1.13.	Smoke carried up by thermals	12
1.14.	Oil refinery smoke accumulating at inversion	13
1.15.	Cloud-base inversion: haze top and cumulus	15
1.16.	Small cumulus sheared over at inversion	15
1.17.	Castellanus cumulus clouds	16
1.18.	Cirrus cloud	18
2.1.	Evaporation of narrow cement works plume	22
2.2.	Black sinuous plume	23
2.3.	Plume levelling out at inversion—Mt. Isa	24
2.4.	Plume carried down to ground in gusts	25
2.5.	Area source of smoke—Edinburgh	39
2.6.	Warm plume rising in smooth wind	44
2.7.	Warm plume in variable wind	44
2.8.	Bifurcation of warm plume	45
2.9.	Thermals from steelworks—Pittsburgh	50
2.10.	Downflow behind lime kiln	52
3.1.	Cloud top at inversion seen from above	58
3.2.	Crepuscular rays in haze below cloud	59

3.3.	Operation 'Chimney Plumes'—inversion penetration	62
3.4.	Pollution trapped below inversion in lee of mountains	63
3.5.	Inversion horizon seen from the air	64
3.6.	Haze stirred up to cloud base	65
3.7.	Haze up to condensation level	66
3.8.	Pollution trapped behind cold front	68
3.9.	Pollution trapped behind shower	68
4.1.	Fog on mountain surface	71
4.2.	Fog draining in katabatic wind	71
4.3.	Valleys filled with fog seen from the air	72
4.4.	Exeter on a clear day	73
4.5.	Exeter in smoke under anticyclonic gloom	73
4.6.	Cement works plume trapped by valley inversion	74
4.7.	Smog top breaking up	75
4.8.	Power station plume emerging from valley smog	75
4.9.	Hot plume levelling out in stable air	76
4.10.	Low level inversion in wide valley	77
4.11.	Inversion rising after sunrise	77
4.12.	Inversion at snow line as pollution top	78
4.13.	Snow line inversion	78
4.14.	Small clouds make anabatic wind visible	79
4.15.	Cloud at snow line inversion in Alps	79
4.16.	Haze among Jura Mts. seen from the air	80
4.17.	Plume impinging on hillside—Wellington, N.Z.	81
4.18.	Plume blanketing mountainside—Tripoli, Lebanon	82
4.19.	'Smog' at Lower Hutt, N.Z.	82
4.20.	Los Angeles smog seen from the air	83
4.21.	Los Angeles-type smog at Santiago, Chile	84
4.22.	Smog at Santiago seen from within	84
5.1.	Cold washed plume descending—Bankside, London	87
5.2.	Bifurcation of water cloud plume—Battersea, London	88
5.3.	Train 'steam'	89
5.4.	Cooling tower cloud on moist day	91
5.5.	'Dark' cement works plume	92
5.6.	White SO_3 plume	93
5.7.	Black carbon 'smoke' plume	94
5.8.	Oil fire black smoke in sunshine	95
5.9.	Continuous cloud from cooling towers	96
5.10.	Contrails and power station thermal	97

5.11.	Steaming lake	98
5.12.	Steaming river made hot by power station	99
5.13.	Red steelworks plume	100
5.14.	Blue domestic smoke	100
5.15.	Salt haze on coast	102
5.16.	Dust devil	103
5.17.	Harmattan-type dust haze	104
5.18.	Blown snow—blizzard	104
5.19.	Sandstorm—Haboob	105
5.20.	Locust swarm	106
6.1.	Separation on mountain side	109
6.2.	Chimney downwash	110
6.3.	Eddies in lee of hill	112
6.4.	Separation on hill crest	113
6.5.	Separation on a wall	116
6.6.	Eddies in street	116
6.7.	Cold inflow model	120
6.8.	Model of penetrating cold inflow	121
6.9.	Ill-advised chimney shape	122
7.1.	Thermal precipitation on a ceiling	130
7.2.	Anomalous blackening of stone	131
7.3.	Blackened 'shadows'	132
7.4.	Blackening of underside of stonework	133
7.5.	Dust deposition on figures	134
7.6.	Differential blackening of similar stones	135
7.7.	Corrosion of power cables	137
8.1.	'Bonfire' smoke in a wind	141
8.2.	'Bonfire' smoke under inversion	142
8.3.	Cottage smoke downwash	142
8.4.	Domestic smoke trapped by low inversion	143
8.5.	Nineteenth century chimneys	144
8.6.	Tall modern chimney	145
8.7.	Effluent from Kowloon power station	146
8.8.	Pollution far from source	148

PREFACE

THE simplest theories about motion in the atmosphere are valid
only for steady winds over very extensive plains under an overcast sky; or, as the mathematician would express it, for a constant
wind over an infinite flat plane of uniform roughness with neutral
stratification. All attempts to progress beyond this to more complicated situations require assumptions which restrict the
applicability of the theory and the measurement of coefficients
by contrived experiment or the observation of nature. As soon as
the theories lose generality they become applicable in such a few
of the very many situations as to become almost useless in
practice. They are, however, very enlightening to study, and the
multitude of complexities of topography, building shapes, wind
fluctuations, and diurnal variations of stratification present us
with practical problems which have to be dealt with individually
and which we can only hope to cope with on the basis of an
understanding of the basic mechanics and physics.

There is no point in making the theories complicated because
they are applicable in only a very few situations. Most of the
results that come out of formal mathematical solution of diffusion
equations using constant diffusion coefficients are obvious
qualitatively, and are not applicable quantitatively in any
practical situation because the assumptions are incorrect. The
purpose of this book is to describe the basic mechanisms whereby
pollution is transported and diffused in the atmosphere, to give
practitioners a correct basis for their decisions. So often have I
met experts using formulae in situations to which they are not
applicable that I may seem to give exaggerated emphasis to the
complexities which make the formulae wrong, and too little
advice on what to do instead. But the complexities are the essence

of it, and make almost every situation unique, which is why we have so many pictures.

Even the photographs which it has been possible to include are a selection from several hundred, each of which has its own different story to tell. The emphasis is upon seeing the mechanisms in action so that one can quickly learn from eye observations what happens with a greater quantitative precision than by theoretical formulae.

The art of observing needs to be practised all the time one is out of doors, for only in this way can enough experience be accumulated to be of any value.

Every scene almost unconsciously analysed for the causes of this whiff of odour in the street and that blackened wall, this town's pall of smoke, that dirty cloud drifting out to sea 1000 metres up. Pollution is part of our scenery; perhaps the most important part of our civilisation which leaves no trace for the geographer and archaeologist.

Even today there is a pretence that it does not exist, just as the spas and resorts represent themselves in the posters as bathed in perpetual sunshine, oil refineries and steelworks, even gold mines and other sources of our wealth, advertise themselves without the pother and stink which always goes with them. But the more aware one becomes of the difficulties and present ignorance, the more interesting does the subject become.

The scientist must not propose facile solutions to social problems. He must become sympathetically aware of the difficulties of legislators, and above all he must recognise that the only solution of air pollution problems is not to make it. Technology alone can facilitate the transition to clean air; meanwhile a fuller understanding of the present situation will make the priorities clear.

Finally, I cannot help hoping that for some the continual observation of air pollution, natural and artificial, will become a real hobby as a result of this collection of pictures.

ACKNOWLEDGEMENTS

The author wishes to thank the following for permission to use their photographs:

1.1.	Charles E. Brown
3.3.	Central Electricity Generating Board
4.20.	A. J. Hagensmit
4.21, 4.22.	J. Stock
5.8.	R. W. Davies
5.9.	G. N. Stone
5.16.	M. P. Garrod
5.18.	P. F. Taylor
5.19.	Air Ministry (Crown Copyright)
5.20.	H. J. Sayer
7.7.	Central Electricity Research Laboratories
8.7.	G. J. Bell

The dust gauge shown in figure 4 was developed at the C.E.R.L. and the drawing is reproduced by permission of the C.E.G.B.

The other pictures were taken by the author.

The experiments shown in plates 6.7 and 6.8 were conducted by O. Jörg.

A number of pictures in this book are included in the pair of colour filmstrip/slide sets entitled "ENVIRONMENT: Air Pollution, Part 1 Local, Continental and Natural Pollution; Part 2 Industrial Pollution". These filmstrip/slide sets, with photographs and lecture notes by R. S. Scorer, issued by Diana Wyllie Ltd., in 1970, are available either from Diana Wyllie Limited, 3 Park Road, Baker Street, London, N.W.1 or from Robert Maxwell and Co. Ltd., Headington Hill Hall, Oxford, at £3 each part.

CHAPTER 1

Pollution over Flat Country

THE STABILITY OF THE AIR

WHEN the wind is light, the agent for removing pollution is thermal convection, which stirs the pollution into an ever-increasing volume of air and brings down to the ground fresh gusts of unpolluted air from above.

Most of the radiation from the ground and from the sun passes through clear air, so that for practical purposes we may regard clean air as transparent to it. Near the ground, that is in the lowest 10 metres or less, there is a perceptible exchange of heat with the ground below, and this merely enhances the effects of conduction from the ground: clean air may be said to be heated or cooled "by proximity to" the ground rather than only by direct contact with it.

By day the sunshine warms the ground, even through an overcast of moderately thin cloud; but all the time the ground loses

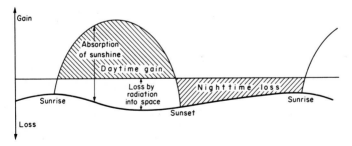

FIG. 1. The diurnal variation of heat received and lost by the ground under clear sky. On the average the net gain by day offsets the net loss by night

1

heat by radiation, and only when the rate of heating by sunshine exceeds this loss does the ground become warmer. When it is sufficiently warm, convection currents begin.

Because it is compressible, rising air expands, since it is moving to lower pressure; as a result of the expansion it is cooled. The amount by which it is cooled may easily be computed from the equations embodying the gas laws, namely

$$p = R\rho T \tag{1}$$

which relates the pressure p to the density ρ and the temperature T by means of the gas constant R, which depends on the chemical nature of the gas;

$$\frac{p}{p_0} = \left(\frac{\rho}{\rho_0}\right)^{\gamma} \tag{2}$$

for adiabatic changes, in which p_0 and ρ_0 are the starting values in a change and γ is the ratio of the specific heats at constant pressure and constant volume of air; and the hydrostatic equation

$$\frac{\partial p}{\partial z} = -g\rho \tag{3}$$

where g is gravity and z is the height.

If a rising parcel of air has the same pressure as its environment, and only an infinitesimally different temperature, and the environment is in hydrostatic equilibrium we can, by eliminating ρ from (1) by means of (2), and then differentiating with respect to z, obtain the following value for the vertical temperature gradient by means of (3):

$$\frac{\partial T}{\partial z} = -\frac{(\gamma - 1)g}{\gamma R} \tag{4}$$

The conditions we have postulated are those of neutral equilibrium, and if the gradient exceeded the value given in (4) the parcel would rise into air cooler than itself and would therefore continue to rise. But if the gradient were less than this, a rising parcel would ascend into air warmer than itself and would therefore sink back again. The air is therefore said to be stably, neutrally, or unstably stratified according as the temperature decreases more slowly, at the same rate, or more rapidly upwards than the value given in (4).

The neutral gradient is called the *dry adiabatic lapse rate;* dry, because the latent heat of condensation of water in clouds is not involved (see page 14). Its value for the atmosphere is constant and is approximately 1°C per 100 metres.

Neutral conditions are not normal, especially in light winds. To establish them the air has to be vigorously stirred, and this can happen either because the wind is strong and eddies are produced mechanically as the air passes over rough ground, or because the ground has been hot and has stirred the air up by thermal convection but has since cooled down a little so that the unstable state no longer exists.

In light winds the ground is usually either cooler than the air, in which case the bottom layers become stably stratified, or it is warmer, and the gradient is unstable near the ground. Figure 2 shows typical temperature gradients at various times of day.

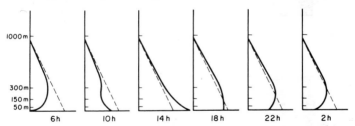

FIG. 2. Typical variations of the temperature in the layers near the ground are shown in successive temperature profiles for a sunny day. The dashed line shows the neutral temperature gradient: its slope is the adiabatic lapse rate

STABLE AIR

It is easy to recognise when the air is stable by the horizontal stratification. Stability at high levels and in valleys is discussed in Chapters 3 and 4. Near the ground it is often manifested by the formation of radiation fog (Plate 1.2), which occurs when the air is cooled below its dew point. Fog is not usually formed if the air is absolutely calm, for then the moisture is deposited as dew on the cold ground or grass; but if there is stirring as a result of a light wind of 2 knots or so, the cooling is spread through a few metres of air and fog is formed because the moisture condenses on to any nuclei that happen to be present in the air, and these are always plentiful. Plate 1.1, from the air, shows how even in the almost flat territory near Norwich, the cold air tends to drain into the areas of lower ground. The fog is as shallow as the trees.

PLATE 1.1. Shallow morning fog at Norwich. The power station chimney protrudes through the very stable layers, but the steam of a train is trapped at a low level

When there is a slight drift of the air, smoke from a cottage or bonfire is carried horizontally at a low level (Plate 1.3). Although it emerges from the chimney much warmer than the air, it mixes rapidly with it so that its excess temperature is very small after it has risen only a metre or two, and then it rises no further because the air is very stable. Plate 1.4 shows a calm morning on which the dark plume from a low chimney marks out the air that has passed over it. Obviously the sunshine which enables the mist and smoke to be seen has already warmed the ground and decreased the stability. Pollution is much more strongly

PLATE 1.2. Fog on a calm August morning at Cézariat, France

PLATE 1.3. Cottage smoke drifting almost undiluted just after sunrise at Much Wenlock. Below it very shallow mist can be seen over the ground which is still in the shadow of the hill and has not been warmed

PLATE 1.4. Fog and smoke at dawn at Dacca airport, showing a long undispersed trail of dark smoke from a small office chimney

PLATE 1.5. A layer of smoke which had drifted across Karachi from the harbour one evening, visible in the light from the President's house illuminated for a state occasion

trapped in the lowest layers of air at night, and Plate 1.5 shows an occasion on which the accumulation was visible.

In summer, before the sea is fully warmed up, tropical air moving to higher latitudes is cooled from below, and often sea fog is formed in this way. Sea fog is fairly shallow because, owing to the smoothness of the sea, only a shallow layer of air is stirred and cooled by it unless the wind is very strong. Plate 1.6 shows sea fog drifting across the coast and causing the air for a few miles inland to be very hazy.

When the air is absolutely calm, dew is deposited on the ground when it is cooled at night. But drops of water on the grass could be produced by guttation, which is the exudation of water from the tips of the blades of grass when the roots are much warmer. Guttation is recognisable as a few large drops originating at the tip; dew takes the form of a multitude of smaller droplets distributed along the blades of grass (Plate 1.8).

UNSTABLE AIR

The temperature of dry sand, bare rock, or a car roof in sunshine, is much higher than that of the air a centimetre or so from it, and at further distances the gradient of temperature decreases. It usually approaches the dry adiabatic lapse rate at 100 metres or so, but at a greater height over a dry desert, and at a lower height over grassland which is kept cold by the evaporation of moisture from the vegetation.

The very large unstable gradient close to the surface is often visible because of the mirages it produces. Rays of light inclined at a glancing angle are reflected upwards as if from pools of water on the road, whereas more steeply inclined rays are not reflected. The gradients needed to produce mirages are of the order of 50 times the dry adiabatic lapse rate (Plate 1.10).

As air moves inland, convection quickly begins on a sunny day and sea fog becomes cumulus cloud (Plate 1.9). An important property of clouds of water droplets is illustrated here: the cloud is not appreciably warmed directly by the sunshine. Sea fog may remain as fog all day over the sea, and at night it is not dispersed

PLATE 1.6. Sea fog drifting across the Welsh coast in July in air of recent tropical origin

PLATE 1.7. Sea fog seen from Cape Cod. It is caused by the advance of warm moist tropical air over the cold water of the Labrador current

PLATE 1.8. Guttation, large drops exuded from the tips of the blades of grass when the ground is warm and the air cool

PLATE 1.9. Sea fog being transformed by convection into cumulus cloud as it drifts inland from the Dovey estuary (Wales)

PLATE 1.10. A mirage is seen on a concrete road when the viewpoint is at 30 cm but not when it is at 150 cm from the ground

over land. By day the sunshine must first penetrate it and warm the ground after which thermal convection warms the air and disperses the fog. The temperature of the sea is changed only very slowly by sunshine.

Thermal convection can be produced artificially, as by the forest fire seen in Plate 1.11 or the power station in Plate 1.12.

If a source of pollution is not warm, the smoke from it may nevertheless be carried upwards in a rather similar manner by natural thermals, and this is illustrated in Plate 1.13. If the source is emitting only a small amount of heat it nevertheless often becomes a preferred region for the ascent of thermals if convection is taking place over the area anyway, and consequently the polluted air has a greater chance of rising than other air, and the chance is increased the greater the amount of heat emitted with it.

PLATE 1.11. Cumulus cloud formed over a forest fire near Vichy. The smoke is carried up to the level at which condensation occurs but no higher

PLATE 1.12. Cumulus formed by the heat from the chimneys and cooling towers of Hams Hall power station between Coventry and Birmingham. The shadows of the cloud on the hazy air below are clearly visible as dark rays

PLATE 1.13. Thermals—bodies of warm air rising from the ground and carrying the smoke, from a low level source, upwards in lumps. This is typical of a day of moderate convection in sunshine, and was photographed at Longton, Staffordshire. The pictures, separated by about a minute, show the smoke first travelling along the ground and then rising in thermals

FUMIGATION

Often the air has a very stable layer at a height of a few hundred metres (see Chapters 3 and 4) and this acts as a lid to the ascent of plumes of pollution from hot sources (see Chapter 2). There is often an accumulation of pollution beneath this stable layer during the night, and during the following morning when the ground becomes warm, thermal convection gradually extends up to it. When the air is thoroughly stirred some of the accumulated pollution is carried down to the ground. This rather sudden increase in low level pollution in the morning is called fumigation.

One of the interesting effects of fumigation is that the pollution is not necessarily observed to follow the wind at the ground. Plate 1.14 shows the plume from an oil refinery reaching up to the condensation level so that a small cumulus cloud is formed. This quickly evaporates and the pollution is carried away with the wind at that level. On this occasion the wind at the surface

PLATE 1.14. The rather smelly plume from an oil refinery at Milford Haven levelling out at a stable layer. Later it was stirred down to the ground by thermal convection and the surface air was "fumigated"

was from the observer towards the refinery, but because the upper wind was in the opposite direction later on, the observer could smell pollution when fumigation carried it down to the ground.

Fumigation occurs everywhere at more or less the same time. Consequently all the points beneath the plume observe pollution at the ground together. If the phenomenon is not understood it is surprising to find the smells arriving before they have been reported at places between the source and the observer.

THE CLOUD-BASE INVERSION

"Inversion" is a term which has come into common use to mean a very stable layer. Originally it referred to layers in which the temperature increased upwards, because, it was said, this was an inversion of the normal state of affairs in which the temperature decreases upwards. But in the neutral state the lapse rate is the dry adiabatic, not zero; and so the term inversion is more commonly used to mean a rather shallow, very stable layer. Deep stable layers are referred to as such, for it is very rare for a deep layer to have the same lapse rate throughout its depth, and it is the positioning of the very stable but shallow layers within it which determines its most important properties.

Thermal convection is a mechanism which stirs the air in such a way as gradually to make its composition uniform. Thus it carries upwards pollution which originates at the ground, and in fumigation carries higher level accumulations downwards. These statements are true of the air below cloud base, but condensation of water into clouds in upcurrents changes the lapse rate (i.e. the vertical temperature gradient) at which the air has zero stability. The latent heat released by condensation is so large that the temperature of a rising parcel of air decreases at much less than the dry adiabatic rate. The actual neutral lapse rate, which is the rate at which a parcel rising without mixing, but condensing cloud, would have the same temperature as its environment, is called the *wet adiabatic lapse rate*. It varies according to how much water is being condensed, and in tropical climates is only about $4\frac{1}{2}°C$ per km while at $0°C$ it is about $7°C$ per km.

PLATE 1.15. The air in between clouds is stably stratified and so the air above cloud base does not become polluted to an extent comparable with the air below. A pronounced haze top is therefore visible at about that level. This view contains north Kent, where there are convection clouds, and part of the Thames estuary, where clouds are absent

PLATE 1.16. A small cumulus formed by the ascent of a thermal through the condensation level being rapidly evaporated in the much drier air above. Its shadow is clearly visible on the haze which is much more dense below the condensation level than above

The air above the condensation level is generally not saturated, and although thermals begin to rise as cumulus clouds when they pass that level, the cloud rapidly evaporates all over its outside and soon disappears if it is not reinforced by succeeding thermals (see Plate 1.16). If the lapse rate were nearer the dry than the wet adiabatic the convection would be very vigorous (see Plate 1.17) and the environment would be warmed rapidly until its lapse rate approached more nearly the wet adiabatic.

The air which descends in between clouds warms up at the dry adiabatic lapse rate because it contains no cloud. If there is any sinking motion to compensate for the thermals which rise through the condensation level the temperature at a fixed point must rise as a result (see Fig. 3). Consequently the ascent of a

PLATE 1.17. Cumulus cloud formed in air in which the lapse rate measurably exceeds the wet adiabatic, and which is therefore very unstable for saturated thermals. Nevertheless, in between the clouds no convection is occurring because the lapse rate is less than the dry adiabatic. This type of cumulus is called castellanus: almost all its buoyancy is derived from latent heat of condensation, and the thermals pass through the condensation level with a very small velocity and excess temperature

few thermals through the cloud base warms the whole air mass and forms a stable layer called the *cloud-base inversion* at that level: only the warmest then rise to form cumulus and the majority are halted.

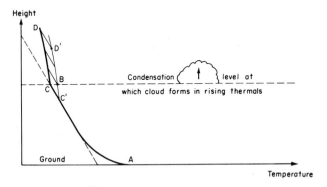

FIG. 3. The line AC represents the temperature profile on a typical day of thermal convection, and the gradient over most of the layer is close to the dry adiabatic lapse rate. If thermals form and their environment sinks bodily as they ascend the air above the condensation level, originally near to the wet adiabatic lapse rate represented by CD, sinks dry adiabatically to C'D'. Consequently the layer C'B is stably stratified although it is below cloud base. This is the cloud-base inversion

The cloud-base inversion is developed more when there is no horizontal convergence of air into the area below cloud base, for then there must exist downcurrents to compensate fully for the upcurrents. But if, for example, there are strong sea breezes blowing inwards horizontally it is possible for there to be no downcurrents between the clouds and no very marked cloud-base inversion is formed. Even then, however, the only air above cloud base which has become polluted is that into which the thermals have ascended as cumulus clouds and evaporated. Their environment is still unstirred and a decrease in pollution is still very noticeable to an observer rising through the cloud base level when the cloud is well broken.

RADIATION FROM CLOUDS

We have already seen that fog can remain over the sea all day because shallow clouds do not absorb an appreciable amount of sunshine which has most of its energy in visible wavelengths. They transmit some and scatter most of the rest back into space. But their own radiation is at infrared wavelengths corresponding to their temperature, and they lose heat to space above very much in the same way as the ground does. When there is a layer of cloud the ground receives almost as much of this long wavelength radiation back from the cloud as it emits itself, because the cloud is only slightly colder, and so radiation fogs do not form under low cloud. Occasionally it forms under a thin layer of high cirrus cloud because the temperature of that cloud could be around 40°C colder than typical low cloud and because it often forms a semi-transparent screen (Plate 1.18).

PLATE 1.18. Cirrus cloud which is often at a temperature near to –40°C does not provide as good a screen as low cloud, which is warmer and more opaque, against loss of heat by radiation from the ground. It delays but does not completely prevent the formation of radiation fog

Layers of low cloud or fog lose as much heat from their tops as the ground does, and this we know is of the same order as the heat gain from sunshine during 24 hours. Cooling the top of a layer of air produces convection as effectively as warming the bottom and so the layer of air below a cloud top becomes well stirred. The heat loss from the cloud top is therefore communicated to the whole layer below it and the temperature drop is much less than if the ground were cooled with only a shallow layer of air close to it. This is a second reason why fogs do not form at night below cloud.

When pollution is emitted into air which does not contain cloud and which is stably stratified by proximity to the ground, it tends to find its own level according to its buoyancy. Bonfire smoke levels out at a few feet above the ground and plumes from power stations at a few hundred feet (see Plates 1.3, 4.10, 8.2). But if it is emitted into a wet fog or into air below low cloud it will probably become mixed into the whole layer. Sometimes this is an advantage, for very low level pollution from houses is carried upwards and away from the ground; but it is a disadvantage if high level pollution from tall chimneys is carried down to the ground. In the latter case a kind of fumigation is taking place, but the transport of the pollution is by means of up and down currents of the order of 1 metre per second or less whereas thermal convection by day from the warm ground produces vertical velocities of ten times this magnitude.

In smog situations when wet fog or cloud is formed in the polluted stagnant air, the whole stagnant mass tends to acquire a uniform composition because of the convection. Polluted air in valleys when there is no cloud has a much more patchy distribution of pollution (see Chapter 4).

CHAPTER 2

Dilution: Formulae and Mechanisms

DISPERSION BY WIND: THE MEANING OF AVERAGES

BEFORE industrial engineers began to build tall chimneys in order to obtain a good draft, architects had realised that it was necessary to make house chimneys protrude some distance above the roof of a house in order to avoid the smell of coal smoke. While the architects seem almost to have forgotten this purpose for chimneys and it has been fashionable to have squat or even hidden chimneys on houses, engineers have in recent years concerned themselves with tall chimneys as a means of avoiding the effects of the pollution they emit. It is a very complicated business to compound the various requirements of performance as a remover of gases from a furnace and as a disposer of pollution, the cost, the appearance, and the structural design, into a single calculation, and so there has always been an insistent demand for a formula to represent each factor. Our purpose here is to explain why it has not been possible to express the performance of a chimney as a polluter of the air in a single formula.

The first fact we have to face is that on some days the emissions from a chimney are quite harmless and are carried away by the wind and upwards by eddies so that they are never detected, while on other days they are very obnoxious although the emission from the chimney top is exactly the same. The great variety of weather which we experience is the cause of this. We cannot hope to predict the concentration of pollution produced at the ground by a chimney when it varies by a factor of perhaps 1000 from one day to another.

If we ask for an average which represents the long-term effect it is important to be clear about what the average is supposed to be. If a concentration of sulphur dioxide of one part per hundred million for one year produced the same effect as 365 parts per hundred million for one day or as 5 parts per thousand for a minute it would be possible to give a sensible definition of the average. But one very concentrated one minute dose may do irrevocable damage to animal or vegetable organisms which have powers to grow new tissue and dissolve chemical substances at such a rate that the same dose over a year has no effect at all. On the other hand there are some effects of sulphur dioxide which are very slowly produced, such as some forms of corrosion and blackening of stonework, and occasional high concentrations do not increase the effect. Chemical effects are influenced by wind speed, humidity, and temperature to such an extent that an average which is correct for one effect may be nonsensical for another. But, before we try to define the average concentration from the point of view of the effects the pollution produces, we must examine the mechanisms in the atmosphere which cause changes in concentration.

THE EFFECT OF WIND SPEED AND GROUND HEATING

"Other things being equal", of which the meaning merits lengthy discussion, the chief effect of wind is to distribute the pollution into a volume of air proportional to the wind speed. The pollution strength P is therefore inversely proportional to wind speed U, or

$$P = \text{const.} \div U. \tag{5}$$

The concentration multiplied by the wind speed integrated over an imaginary vertical surface placed across the track of the air gives the rate at which pollution is carried away, and must be the same as the rate of emission. If equation (5) is to be strictly true, always with the same value of the constant, the area of our imaginary surface which is polluted must be the same in all cases. Therefore, the spread must be the same at a given distance in all wind speeds which means that the eddy velocities which cause the spreading must be proportional to the wind speed, or to put it another way, the eddy shape must be the same in all speeds, and not more elongated in high winds. This is one of the "other things" which must be "equal".

There are several possible causes of eddies. They may be produced mechanically by the wind flow over obstacles and ground roughnesses, in which case they are very likely to be of a shape determined by the obstacles and will be the same at all but very small wind speeds. But if they are caused by thermal convection their intensity depends on the rate at which sunshine heats the air by warming the ground, and this is only slightly affected by wind speed. Generally in light winds the thermally produced eddies are larger than in strong winds when the mechanically

PLATE 2.1. The plume from Hope cement works in a Derbyshire valley on a day of strong wind but very stable air. The ground was cold with some snow-covered patches. The eddies at the height of this chimney top were feeble so that the spread of the plume is very small.

The whiteness of the plume is due to the condensed water vapour which forms a cloud sometimes inside the chimney and sometimes not until a short distance from the chimney top (see Chaper 5). As a result of dilution by eddies the water droplets evaporate leaving only a faint plume of dust particles further downwind

produced eddies are more effective in transferring heat to the air above. The consequence of this is that a plume of pollution from a chimney is usually spread over a greater area in light winds than in strong winds in sunshine, but when the air is slightly stabilised by cold ground below there is very little spread in light winds, but a strong wind is almost as turbulent as when there is sunshine. These features are illustrated in Plates 2.1, 2.2, and 2.3.

PLATE 2.2. The black plume of Poolsbrook colliery in Derbyshire showing a sinuous form in the large eddies produced by thermal convection from the warm ground in sunshine. The smoke does not travel along a sinuous track but the sinuosities travel along with the wind growing in size as they go

The conditions illustrated in these three pictures are so different from one another and from the situations shown in Plates 2.4, 4.17, and 5.9, that we must either abandon the relationship (5) altogether, or say that the constant of proportionality is different in different cases, or that the pollution referred to is only some sort of long-term average and that U is an average wind speed. The error of this last way out is that places that have different average wind speeds do not usually have the same frequency of the different angles of spread so that the constants of proportion-

PLATE 2.3. On an almost calm morning at Mount Isa, Queensland, there are no eddies to dilute this plume as it drifts slowly downwind. In the distance the accumulated pollution rests at a few hundred feet waiting to be brought down to the ground by fumigation later in the morning. It was carried in a semicircular arc by the light wind during the hour or two before the picture was taken

ality must be different in different places, and that if they have the same average conditions we cannot have occasion to use the formula which refers to different wind speeds.

Returning to the eddies produced by ground unevennesses, we note that wind direction can also play a part in determining the constant of proportionality in (5). A chimney downwind of a large steep hill or tall building may experience very large eddies such as that shown in Plate 2.4, but when the wind changes direction the air motion downwind from the chimney contains smaller and less intense eddies like those in Plate 2.2.

The meaning of an average may also be affected by the kind of instrument used to measure the pollution. If we wished to measure the amount of dust coming through an open window we ought to let it be carried through naturally by the wind, and for a given concentration of dust in the air the amount collected would be roughly proportional to the wind speed. But if we required to know how much passed downwards through a hole facing upwards we should design an instrument to collect dust in the same way. The eddies in the air obviously have a big effect on the deposition, but apart from their effect the deposition of dust should, for this purpose, be measured like rainfall, and a container exposed with an open hole at the top.

We must not delude ourselves, however, that dust deposition is as easy to measure as rainfall, because it also depends on the nature of the surface. A sticky surface will collect more than a shiny surface from which the wind removes dust easily. Carey suggested a kind of negative method for measuring deposits: he placed several rubber bungs on a horizontal sheet of white paper and removed one each day, and this enabled him to estimate the dinginess caused by the deposits of the last 1, 2, 3, 4, . . ., days.

If we wish to measure the concentration of pollution in the air we have to extract it from a measured volume of air. The measurement made is then independent of the wind speed and of the mechanisms by which the pollution causes undesirable effects,

PLATE 2.4. Ironbridge power station plume (see Plate 4.6) carried down to the ground by large eddies on a sunny day in a hilly region

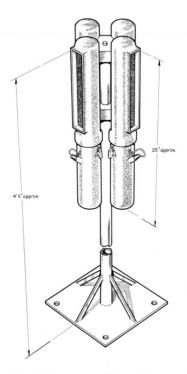

25″approx.

4′6″approx.

Fig. 4. Directional dust gauge designed for use by the Central Electricity Generating Board in pollution surveys around power stations. The correlation between wind direction and amount of airborne dust is not lost whereas a non-directional gauge would have to be read very frequently and very small amounts measured if this information were to be obtained

and to this extent may be called an 'objective' measurement, but this does not mean that it will tell us all we wish to know. It is fallacious, for example, to assume that if the average concentration and the average wind speed are both measured to be the same on two different days that the effects will be the same on the two days. For example all the high concentrations might have occurred at high wind speed on one day and at low wind speed on the other.

The averages can be even more misleading, if not correctly interpreted, in the case of wind direction. In the averaging process information about some very important features of the situation may be lost. For example, it is impossible to deduce from a table of frequencies of different wind directions and a table of frequencies of wind speeds how frequent light or strong winds are from within a given sector. The following example illustrates this point:

We wish to know how frequently the pollution from a new chemical plant will be carried towards the south-east in light winds because there exists a small town to the south-east of the site and only in light winds will the pollution be concentrated enough to concern us. The annual percentage frequencies of winds in excess of 4 knots from directions in various sectors at a nearby airport is given in Table 1.

TABLE 1

direction, °true	350 –010	020 –040	050 –070	080 –100	110 –130	140 –160	170 –190	200 –220	230 –250	260 –280	290 –310	320 –340	Calm	1–3 knots
% frequency	4·7	2·4	4·9	4·2	5·2	10·5	9·8	5·2	7·6	6·5	13·4	9·1	4·8	11·6

Table 1 indicates that winds from the north-west are the most frequent. Evidently there is a problem. The percentage frequencies of winds of various strengths is given in Table 2 and this

TABLE 2

strength, knots	0	1–3	4–6	7–10	11–16	17–21	22–27	>27
% frequency	4·8	11·6	14·4	23·4	29·3	10·8	4·4	1·3

shows that winds below 7 knots occur for a substantial part of the time. The worst conditions are those in which fog is likely to form, and so we obtain information indicating the frequency of fogs and find that they occur predominantly in the morning and in winter and amount to a total of 120 hours per year at the airport, with a visibility less than 200 metres.

Before we can draw any conclusions from this information we need to know whether the frequency in winter of winds from the north-westerly sector is greater or less than for the whole year, because in winter stable conditions are more common. We also need to know whether the light winds have the same distribution in direction as all winds. When this investigation was carried out it was found that in the three winter months the percentage frequency of winds in the range 4–6 knots from various directions was as given in Table 3.

TABLE 3

direction	350 −010	020 −040	050 −070	080 −100	110 −130	140 −160	170 −190	200 −220	230 −250	260 −280	290 −310	320 −340
% frequency	1·1	1·1	0·4	0·8	1·3	0·8	1·6	1·3	0·7	0·7	0·5	1·1

Experience of the drift of fog and smoke in winds of 3 knots or less showed that it nearly always drifted from the south or south-east. This showed that the trend indicated by Tables 1 and 3 for north-westerly winds to be less frequent at low speeds was continued. The most common wind was from 290–310 in the range 11–16 knots, which blew 11% of the time in July; but since this was predominantly a summer, daytime, fresh breeze it was of little concern.

Thus at first it seemed, since more than a quarter of the time the wind blew from within the north-westerly sector, that more than a quarter of the 120 hours of fog, and of the many hours of similar conditions nearly producing fog, would occur with a drift of air from the new plant towards the town. However, it turned out that the north-westerly winds were predominantly strong summer winds, and that when the wind was light and the air very stable the direction was almost always different, and generally from the south-east.

This investigation was concerned with conditions in which gaseous pollution might cause a nuisance. Had the plant been a source of grit and dust instead, the situation would have been quite different because only in strong gusty winds, such as those which are common from the north-west in this area in summer would the dust have been carried as far as the town in appreciable quantities. The situation would have been completely reversed had the town been situated to the north-west of the plant, for then the drift of air from the plant in stable conditions and light winds would have been towards it.

It is essential, therefore, to obtain observations of all the required combinations of factors in order to estimate the frequency of occurrence of potentially objectionable circumstances. In the above example the situation was unusual because the topography makes winds from the north-west and south-east much more common than from other directions. It was this fact which made the danger from north-westerly winds seem greater than it actually was. But it should be noted that the topography has a much greater effect on the direction of light winds in stable conditions than on almost any other kind of wind so that in an area with seemingly well-distributed winds the conditions worst for pollution may produce a drift of the air in one direction far more often than in any other.

CONICAL AND PARABOLIC PLUMES

If the diffusion of a smoke plume is caused entirely by eddies in the atmosphere which are small compared with the width of the plume the spreading is of a similar form to, but much more rapid than, spreading by molecular diffusion. The sideways, upwards and downwards transport of material is proportional to the gradient of concentration which in turn is inversely proportional to the plume width. This means that the rate of change of the plume width, r, is proportional to r^{-1}, and so

$$r \propto t^{\frac{1}{2}} \tag{6}$$

where t is the time from release into the air. Since the horizontal travel is more or less with the wind, U, the distance x downwind is given by

$$x = Ut \tag{7}$$

and so

$$r \propto x^{\frac{1}{2}} \tag{8}$$

In calculating this kind of diffusion (called Fickian) the plume width has to be defined by the contour containing, say, 90% of the material in any section transverse to the wind, or the contour on which the concentration is, say, 5% of the maximum. The equation giving the flux of the pollution is

flux per unit area $= K \times$ gradient of concentration   (9)

and this implies that the plume has no precise edge but fades gradually to nothing at infinite distance. The error is not important provided that we define the theoretical edge in the manner just described, but in practice a plume does have a quite definite edge beyond which no pollution belonging to it is present.

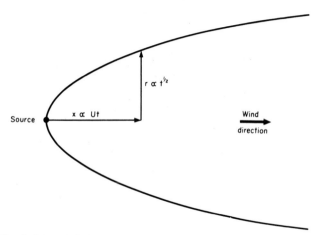

FIG. 5. If the diffusion is the same as that due to a constant diffusion coefficient, the plume widens along a parabola. The width is defined as the width within the contour on which the concentration is, say, 5% of the maximum on the axis which points directly downwind

Fickian diffusion produces a plume with a parabolic section (Fig. 5). In practice, as we shall see, this is modified by the fact that the turbulence decreases with height so that the value of K for upward and downward diffusion decreases upwards. The value for sideways diffusion is not necessarily the same, and often the most effective agent for sideways diffusion is wind shear across the plume axis. Thus pollution may be carried to a point not directly downwind of the source by being first carried upwards, then sideways at a higher level by a cross-component of wind, and then down again, more effectively than by diffusion by eddies across the wind at a constant height.

The chief objection to this kind of formula is that it is not realistic, not mainly because it gives no well-defined edge to a plume, but because there are many different eddy sizes present in the atmosphere. We have to recognise that all the time we are talking about an average concentration over an as yet undetermined period of time. At points close to the source an average, in the sense that the measurement may be repeated and roughly the same value obtained the second time, can be obtained in 2 or 3 minutes, whereas at a distance of, say, two miles, the fluctuations are of such a long period that the sampling time has to be increased to more like 20 minutes before a repeatable measurement is obtained.

The further we go from the source, the more we find ourselves concerned with the large eddies and less with the small ones. This is because when the plume has a small width the large eddies carry it about bodily and so we can make meaningful measurements on it in a fraction of the time required for a large eddy to pass: we regard the large eddies as being the cause of changes in wind direction. On the other hand, when the plume width is large, the effect of the small eddies has become unimportant for they correspond to a small value of K in the Fickian system of ideas: although they were causing a widening of the plume at a large angle close to the source, at greater distances the parabola to which they correspond is widening at a much smaller angle. The parabola corresponding to the eddies which we thought of as producing wind direction changes near to the source now comes into its own, and the widening still continues at the original wide angle.

The eddies which are important at any given distance from the source are those whose size is comparable with the plume width, and if all sizes were equally present it would be possible

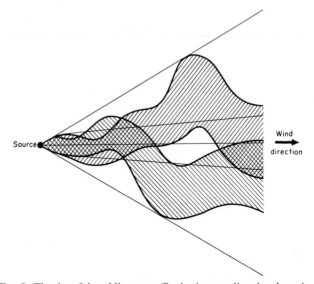

Fig. 6. The size of the eddies most effective in spreading the plume is about the same as the plume dimension. Smaller eddies cause a very slight spreading of the plume; larger eddies can be said to cause changes in wind direction. Eddies of intermediate size cause sinuosities which spread the pollution within a boundary which is roughly a cone

Two positions of the plume are shown. The sampling time required to ensure the passage of several sinuosities, enough to give a representative average, increases with distance from the source and with the size of the sinuosities

for the plume to grow in size along a cone (Fig. 6). Unless we knew in some detail how intense the eddies of different sizes were we could not predict any significant departure from spreading along a cone.

In practice, on a day when the eddies are caused by ground irregularities, buildings, and other obstacles, together with a certain amount of stirring due to thermal convection, the plume has become so dilute by the time its dimensions are large compared with the eddy size that we can, for practical purposes, think of it as spreading along a cone. But we must remember that if we take the same sampling time at all distances we shall obtain

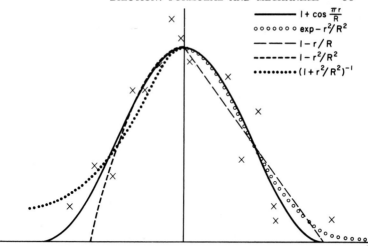

Fig. 7. Profiles of average pollution concentration across a plume according to some simple well known formulae, and typical observations to which they have to be matched marked by crosses. The value of R and the constant of proportionality are adjustable in all cases and they are shown with the same maximum value of P. The vertical co-ordinate is P, the pollution concentration and the horizontal one is R, distance from the centre. The formula corresponding to each curve is given in the key. Only the first shape is shown on both sides of the axis. The others are shown on one side only

a parabolic shape because we are allowing the "changes in wind direction" close to the source to increase the diffusion there. But as we go still further away from the source we shall find that the parabolic outline has become contorted by the larger eddies and is snaking around within the cone. The conical outline corresponds to what we see: it is an envelope of all the sinuosities visible at one moment. To obtain the form of the plume out to any distance it is necessary for the source to have emitted for the time required for the wind to carry material to that distance, which implies a correspondingly large sampling time at large distances.

The question therefore arises how we should represent the spreading quantitatively. The plume of parabolic section is represented by the coefficient K, and the average distribution across the plume is necessarily gaussian. Such a distribution is as good as almost any other for comparison with observation; but it has no special merit because observations always show great irregularities and variations, and it is pointless to pretend to any precision either in measurement or prediction of the distribution across the plume. It cannot be claimed that observations indicate that the concentration profile takes any particular one of the several forms which have been proposed rather than any other. An approximate guess cannot be far wrong: all that is required is that the total flux of pollution along the plume should be the same at all distances and equal to the source strength, that the spread should be about right, and that the profile should be more or less bell-shaped (Fig. 7).

It is a practical impossibility to measure the concentration in a vertical plane across a plume with sufficient precision to discriminate between many proposed formulae. Consequently it is always possible for an author to obtain, from existing measurements, an estimate of the coefficients in his formula. In most cases the measurements are made only on the ground and so it is necessary to include in the theory which is the basis for a formula some assumption about the effect of the ground on the dispersion. The simplest and most commonly made assumption is that the ground reflects the pollution so that the concentration at all points above ground is the same as if there were an image source

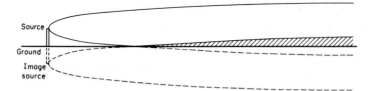

Fig. 8. Source and image source below ground used to calculate the concentration due to a plume (which is shown as parabolic in this illustration). The concentration at ground level is twice that which would be observed at the same point if the ground were not a barrier. In the shaded region, the pollution is increased by the amount there corresponding to the image source

below ground and the ground did not exist. This simply means that the pollution at ground level has twice the concentration it would have had if the ground had not been there (Fig. 8).

No other method has been proposed for representing the ground because any criticism that can be made of the image source method is equally a criticism of all the theories. There are many forms of pollution (for example iodine and fluorine compounds) which are absorbed by vegetation or become attached to solid surfaces. The ground is therefore not a complete barrier to downward flow. The chief difficulties about the ground, however, really stem from the fact that the pattern of motion changes significantly as the ground is approached. At heights of hundreds or even tens of metres the eddies are often nearly isotropic (the vertical and horizontal fluctuations of velocity are about equal), but near the ground the vertical fluctuations in the larger eddies are noticeably reduced so that the dispersion in a vertical direction is correspondingly less.

The simplest way in which a plume can be represented is by means of a cone, the angle of which is used to represent the rate of dispersion. The distribution across a transverse section of the cone may be taken to be according to any simple formula such as $(1 + \cos \pi r/R)$, $(1 - r^2/R)$, or $(1 - r/R)$, for $r < R$, and zero for $r > R$; or we could use $(1 + r^2/R^2)^{-1}$ or $\exp(-\mu^2/R^2)$, where r now goes to infinity. In all these R is a measure of the plume width and is a function of distance downwind, and the distribution chosen has to be multiplied by a suitable constant so that the total transport across a vertical plane is equal to the source strength. For example, if the first formula is chosen and the plume is assumed to have a circular cross-section of radius R, and P is the pollution concentration and U the wind speed, a source emitting a quantity Q of pollution in unit time produces a downwind flux equal to

$$Q = \int UP \, dS \qquad (10)$$

where the integration is over the whole area of a transverse section of the plume. Thus the indestructibility of the pollution is represented in this case (see Fig. 9) by

$$Q = \int_0^R UP_{max} \frac{1}{2}\left(1 + \cos \frac{\pi r}{R}\right) 2\pi r \, d \tag{11}$$

from which

$$P_{max} = \frac{Q}{\pi R^2 U \left(\dfrac{1}{2} - \dfrac{2}{\pi^2}\right)}, \tag{12}$$

and

$$P = \frac{1}{2} P_{max}\left(1 + \cos \frac{\pi r}{R}\right) \qquad r < R$$
$$= 0 \qquad\qquad\qquad r > R \tag{13}$$

Modifications can readily be made to formulae of this kind for cases in which the sideways spread is greater than upwards and downwards. If the upwards spread is greater than downwards, the axis of the cone may be tilted upwards a little. The angle of spread itself is determined by stating what R is as a function of x, the distance downwind, for example $R = x \tan \alpha$ for the conical plume.

Whatever the form chosen for R as a function of x, if its axis is horizontal the plume may be said to impinge on the ground where $R = h$, the height of the source above ground. The concentration at the ground when the contour $r = \lambda R$ reaches the ground, which is when $\lambda R = h$, is proportional to R^{-2} and therefore to h^{-2}. Since this is true for all λ the maximum concentration at the ground always occurs for the same λ, for a given distribution across the plume, for any value of h. The maximum concentration on the ground is therefore proportional to h^{-2}, but the distance at which it occurs and the area over which the concentration is above a certain fraction of the maximum do depend on the distribution across the plume and on the form of R as a function of x. Therefore

$$P_{\text{ground max}} \propto Q/Uh^2 \tag{14}$$

in all cases. This result may be used for purposes of comparing the pollution due to sources of different strengths at different

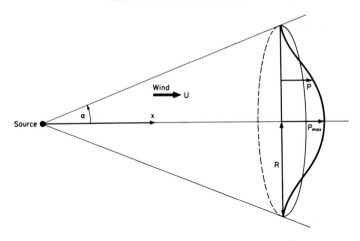

Fig. 9. The simple conical model of a diverging plume. The transverse section is imagined to be circular and the pollution has a maximum in the middle and falls, according to the first formula, to zero at radius R and beyond. The value of the angle of spread, α, has to be chosen to accord with observations of the kind of situation the model is representing, and P_{max} is chosen by equation (12)

heights in winds of different speeds without serious error except when the conditions make all the formulae obviously incorrect. The pollution referred to in equation (14) is the maximum averaged over a specified sampling time; and the sampling time determines the constant of proportionality, other things being equal.

The chief justification of a formula of this kind is simplicity. The means by which it is made to accord with any observations available are clear to the user, and it contains no suggestion that it has greater validity than the simple assumptions suggest. Other formulae can only claim greater validity for special situations in which the particular assumptions on which they are based are known to be rather accurate. Normally the variation in the observations is so great as not to provide for any discrimination between formulae, and this fact alone is an important justification of the simplest formula.

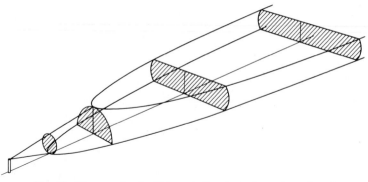

Fig. 10. When confined within a shallow layer by an inversion, the area of the cross section becomes proportional to the width R, and no longer to R^2, regardless of how R increases with distance downwind

When measurements are made a sampling time has to be chosen. In making a choice in favour of say 3 minutes or 1 hour we automatically fix our attention on eddies of corresponding size. The average concentrations measured over 3 minutes will be greater than the 60 minute averages, but spread over a smaller width. The plume, as spread by eddies of period considerably less than 3 minutes, is narrower than the plume as spread by the eddies which take a large fraction of an hour to pass. If we choose any two significantly different sampling times the plume as seen by the longer time averages will be like the effect of the plume of the shorter sampling time meandering about. But in practice there is always an appropriate sampling time for any distance, and it is equal to about the time taken for the wind to carry a particle from the source to the point of observation. Samples taken over shorter periods will not be found to have the maximum in more or less the same place, and samples taken over much longer periods show a gradual increase in the spread as the time is increased, and this continues indefinitely: the meaningful time is the shortest one which does not show a significant shift in the centre of the distribution from one sample to the next, and for measuring the maximum ground level concentration this is usually about the time taken for the wind to carry the air a distance of about fifteen times the height of the source above the ground.

All formulae of this kind are inapplicable to occasions of light wind.

PLUMES CONFINED IN LAYERS

The upward spread of a plume is often halted by the presence of an inversion, that is a layer of great stability which acts as a lid to the turbulence. The effluent is then confined in a layer of constant thickness and soon the dilution is proportional to R^{-1} instead of R^{-2}, provided that the sideways spread continues (Fig. 10).

At considerable distances from the source the appropriate sampling time for the plume may not be the appropriate one from the point of view of the effects of the pollution. The occasional whiffs as the sinuous plume passes over a point may be the only important aspect of a plume when the average over the longer sampling time has fallen below a certain threshold. In that case

PLATE 2.5. Edinburgh seen from Salisbury Craggs. The houses burning coal create an area source of smoke

a two-dimensional parabolic form of dilution is most appropriate so that the cross-section area of the plume and therefore also the concentration, is proportional to R^{-1}, i.e. to $x^{-\frac{1}{2}}$.

AREA SOURCES

At great distances we are not really concerned with the pollution from a single source, because it is already too dilute, but with pollution from industrial areas, housing estates, or even towns and conurbations. When there is a substantial barrier to upward dispersion and the source is wider than the height of the barrier the dilution is proportional to x^{-1}, but in this case x has to be measured from a point some distance upwind of the source, and the dilution is therefore rather slow (Fig. 11). In practice there is a considerable absorption of pollution by the vegetation as the wind carries it away from a town so that when there is an inversion the pollution is often not as bad as would be calculated on the assumption that it all remains in the air. The sea, however, is not as absorbent, and pollution from the Lancashire area is often carried to the Isle of Man, a distance of 80 miles, with very little dilution because under an inversion, such as commonly occurs in

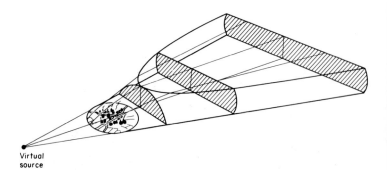

Virtual
source

Fig. 11. The apex, or virtual source, of the wedge or cone down which pollution from an area source is spreading is far upwind of the source, and the dilution downwind at distances from the source up to one or two times the distance from the apex to the source is very slow

east winds in winter, there is very little sideways dispersion over the sea. Because there are no large eddies the dilution, at best, is proportional to R^{-1} which is proportional to $x^{-\frac{1}{2}}$ in the parabolic regions, and from a large area source such as Lancashire the distance x must be measured from, perhaps 100 miles upwind. Consequently the concentration in travelling to the Isle of Man is multiplied by a factor of about $2^{-\frac{1}{2}}$, which corresponds to rather small dilution.

HOT PLUMES: EQUILIBRIUM LEVEL, BIFURCATION, PLUME RISE

Even if the air is calm or the wind free from turbulence the pollution from a source becomes diluted into the air around it if it is warmer or cooler. In Plate 2.3 the warm gases can be seen mixing by means of eddies whose size is rather less than the plume width for a distance of about five chimney heights, after which no further mixing takes place. The mixing ceases when the gases reach their equilibrium level in surroundings which are stably stratified. The same can be seen in Plate 4.9, which illustrates a case of very light wind and a very pronounced equilibrium level. Such plumes sometimes extend for many miles without any appreciable widening. Plate 1.4 shows the same thing on a smaller scale.

In a neutral environment hot gases rise and mix as they do so. If there is no wind a steady source produces a conical plume. The concentration σ and the buoyancy are proportional if there is only negligible heat loss by radiation, and so the upward flux of momentum increases at a rate proportional to the buoyancy force and therefore proportional to the concentration and the area of a horizontal section, i.e. if w is a measure of the vertical velocity, and R of the width,

$$\frac{d}{dz} w^2 R^2 \propto \sigma R^2 \tag{15}$$

The upward flux of material is constant so that

$$\frac{d}{dz} \sigma w R^2 = 0 \tag{16}$$

These equations are satisfied by
$$R \propto z, \qquad w \propto z^{-1/3}, \qquad \sigma \propto z^{-5/3}. \qquad (17)$$
A full justification of this simple result is given in *Natural Aerodynamics*, p 188.

When a smooth cross wind blows, the plume is bent over and rises like a tube of warm gas. In that case also the pollution concentration and temperature excess are proportional to one another and when multiplied by the area of a vertical section are proportional to the rate of increase of upward momentum. Thus in this case

$$w \frac{d}{dz} wR^2 \propto \sigma R^2 \qquad (18)$$

Since the pollution in a vertical section is conserved

$$\frac{d}{dz} \sigma R^2 = 0, \qquad (19)$$

and these equations are satisfied by
$$R \propto z, \qquad w \propto z^{-1/2}, \qquad \sigma \propto z^{-2}. \qquad (20)$$
(*loc. cit.*, p. 195).

The ratio of R to z can be determined by experiment, and, if R is half the overall width of the plume approximately, $R = z/5$ for the vertical plume and $R = z/2 \cdot 25$ for the bent-over plume, and in this latter case a bent-over plume looked at from upwind would appear to be wedge-shaped (Fig. 12). The constants of proportionality in the other equations (17) and (20) are determined by the conditions of each particular case. Thus if w_1 is the upward velocity at height z_1, in a bent over plume we can write

$$\frac{w}{w_1} = \left(\frac{z}{z_1}\right)^{-\frac{1}{2}} \qquad (21)$$

with a similarly derived equation for σ.

The mixing is rapid so that the plume very quickly takes on the horizontal velocity of the environment. Thus for the bent over plume x, the horizontal distance travelled, is equal to Ut after time t, while

$$w = \frac{dz}{dt} \propto z^{-\frac{1}{2}} \tag{22}$$

which implies that

$$z \propto t^{2/3} \propto x^{2/3} \tag{23}$$

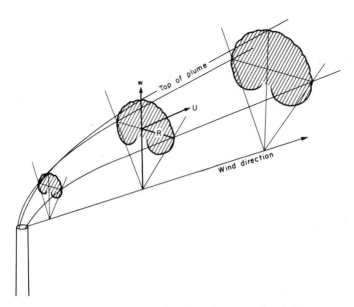

FIG. 12. The width of a bent-over hot plume in a neutral turbulence-free environment is proportional to its height above an origin close to the chimney top. The diagram shows three similar sections transverse to the wind, the size being proportional to height above the origin

The plume in Plate 2.3 starts along a shape $z^3 \propto x^2$, but soon levels out at its equilibrium height. In Plate 2.6 is seen a pair of plumes from two adjacent chimneys. Their rate of rise does not decrease as rapidly as the formula suggests because they amalgamate: they mix with one another and are therefore diluted more slowly than if they were alone. A view from beneath these plumes is seen in Plate 2.8 where the bifurcation due to the internal

PLATE 2.6. The plumes from Keadby power station rising and widening under the influence of their buoyancy on an overcast day with no thermal convection and a fairly smooth wind. The same plumes are seen from below in Plate 2.8

PLATE 2.7. The plume of Kirton-in-Lindsey cement works rising under its own buoyancy in a wind which is smooth except for a few large eddies which cause a variation in the wind strength

circulation induced by the buoyancy is clearly seen. This bifurcation is always clearly seen in hot plumes rising in a very smooth wind (see also Plate 1.1).

Occasionally, for example when the sun breaks through clouds and quickly begins to create thermal convection in an otherwise rather smooth wind, there are only large eddies present. In that case the strength of the wind appears to vary and a plume widening according to equations (20) and (23) becomes sinuous. This is illustrated in Plate 2.7.

Evidently a hot plume in a neutral environment would continue to rise indefinitely if there were no other eddies, but observation of actual hot plumes shows that the bifurcation is not usually visible because of the eddies in the surrounding air. Since the velocity of rise and the velocities in the eddies induced by buoyancy are proportional to $z^{-\frac{1}{2}}$, i.e. to $t^{-\frac{1}{2}}$, the eddies in the environment will become dominant after a finite time, and then the plume will behave as if it had no buoyancy. The idea has

PLATE 2.8. The plumes of Plate 2.6 seen from below. The buoyancy causes a bifurcation which is always seen in a smooth wind but which is often disrupted by more intense eddies in the surrounding air

been widely favoured that when this stage is reached the plume may be thought of as having been emitted from a source whose height is greater than the actual source's by the amount the plume rose while the buoyancy was effective. Several methods of computing the so-called thermal rise, have been proposed and are widely used, but they all fail in fact because it is not possible to express them correctly in a useful form.

Perhaps the simplest approach illustrates the difficulties best. The buoyancy per unit length of plume is proportional to U^{-1}, and if Q is the quantity of buoyancy (i.e. volume $\times$ fractional absolute temperature excess) emitted in unit time, a more complete version of the second of equations (20) is

$$ w \propto \left(\frac{gQ}{Uz}\right)^{\frac{1}{2}} \tag{24} $$

where g is gravity. The plume will rise no further but will be passively diffused when w has decreased to a certain multiple of the velocity of the environmental eddies. Thus if v_e is the typical eddy velocity the thermal rise h' would be given by

$$ v_e = \mu\left(\frac{gQ}{Uh'}\right)^{\frac{1}{2}} \tag{25} $$

where μ is a multiplier that has to be determined by observation on at least one occasion, and which should be a universal constant. This equation shows that the thermal rise, h', depends not only on the buoyancy emitted, Q, but also on U^{-1} and v_e^{-2}. Any other formulation of the problem, however complicated, must show that h' depends on U and v_e in some way. Some derivations give different powers of U, v_e, and Q in the formula either because the plume rise has been incorrectly defined as the height reached at a certain distance downwind or the height attained at a certain time after leaving the chimney on the grounds that this fits in better with the methods used for making observations of actual plumes, or because the plume is assumed to be rising according to a formula different from (20) because its geometrical shape has been assumed to be different (for example a series of discrete lumps instead of a continuous stream). But as long as they approach the problem in the same simple way as the derivation of formula (25), that is to say they assume that the changeover from buoyant rise to passive diffusion by ambient eddies can be represented as being more or less sudden, the chief term in the equation must be as given above.

The only alternative is to assume that the ambient eddies cause a significant dilution from the start so that the two mechanisms of mixing are both operating all the time. The difficulty with such a theoretical approach is that there is at present no theory to tell us what the buoyancy will do when the plume is diluted by ambient eddies, and it is necessary to fall back on observations or experiments: since such experiments are in fact a direct observation of the end phenomenon which the theory is about, the theory is redundant. The major problem of actually observing the moment when the plume rise is complete remains, and has not really been solved because even when it is complete the top continues to rise, before it is complete the bottom begins to descend, and when the rise is complete the dilution is usually so great that the plume cannot be accurately observed from the ground anyway.

But let us suppose that all these problems were solved and that a reliable formula of the form

$$h' = KQ^\alpha U^\beta v_e{}^\gamma \tag{26}$$

(or some other function of the same quantities) had been found. It is evident that the value of h' is still very much a feature of the environment (which determines U and v_e) and is not simply a property of the chimney and its output.

It has often been argued that

$$v_e = \lambda U \tag{27}$$

because the eddies are caused by the wind blowing over obstacles and roughnesses, λ being a constant depending on the nature of the obstacles and roughnesses. If this is combined with (25) we find that for a given chimney output

$$h' = k\lambda^{-2}U^{-3} \tag{28}$$

Since, also, the effective stack height H is the actual height h increased by h', according to (14), for a given output

$$P_{\text{ground max}} \propto U^{-1}(h + k\lambda^{-2}U^{-3})^{-2} \qquad (29)$$

and this has a maximum when U is such that

$$h' = \frac{1}{5} h \qquad (30)$$

On the other hand, if we assume that the eddies in the air are given regardless of the wind speed, v_e may be taken as a constant and then

$$P_{\text{ground max}} \propto U^{-1}(h + KQ^a v_e^\gamma U^{-1})^{-2} \qquad (31)$$

which has a maximum when U is such that

$$h' = h \qquad (32)$$

The object of computing the wind speed at which the ground level pollution is a maximum has been to simplify the calculation of $P_{\text{ground max}}$ for the ordinary user of the formula, by substituting this value of the wind speed in the formula from the start. This method is used by Wippermann and Klug (*Int. J. Air Wat. Poll.* **6**, p.27), but it takes no account of the frequency of occurrence of winds of the speed so computed, and is in dispute anyway because formula (26) is not agreed and so (30) or (32) or some intermediate form cannot be decided upon.

One of the chief reasons why (26) is uncertain is that v_e varies enormously from day to day and even from hour to hour, and whether the form of (26) is agreed or not, even if the values of α, β, and λ are undisputed, v_e is necessarily of major importance in any formula.

The great variation in v_e, the ambient eddy velocity, can be thought of as a great variation in the value of λ in (27), which relates it to the wind speed. During a sunny afternoon when the wind is light the eddies are mainly thermal convection and the wind gusts may be as great as the average wind, and so $\lambda = 1$. In the evening, when the ground becomes cold and the air close to it stable, tall chimneys emit plumes at a higher level into the air which has not yet been cooled and is therefore in a state of neutral stratification. The great stability close to the ground is accompanied by a decrease of wind there and a damping out of the eddies so that for a time before the cooling reaches the chimney top level, the plume is emitted into a turbulence-free neutrally-stratified environment, and rises indefinitely (i.e. to the cloud base level) according to equation (23) and shows a very clear bifurcation. At this time we would say that λ is of the order of $0\cdot01$ or less.

It is evident from (28), or (26) and (27), that such a change in λ alters by a large factor the wind speed U at which (32), or (30) if that is preferred, is satisfied. It is also well known from experience that during an afternoon with strong turbulence the effluent is frequently brought down to the ground, unless the buoyancy is very large, fairly close to the chimney, in the manner shown in Plate 2.4 (where the gustiness was enhanced by local hills). In the evening, however, the pollution stays aloft when the air below the chimney top becomes stabilised, and this marked evening decrease of ground level pollution from high sources is very noticeable near brickworks where there are many chimneys, none having a very large heat output. At the same time pollution from low level sources may be trapped at a low level and produce large concentrations.

If a formula were devised to give a thermal rise it would still only be applicable when the stratification of the air was uniform, probably neutral. A formula for a uniform non-neutral stratification is rather useless because a non-neutral stratification is scarcely ever uniform. The stratification, whether stable or unstable, is usually much greater near the ground because it is usually caused by a change of ground temperature, and the variations with height are of major importance.

It should be evident to the reader that the study of the attractive concept of thermal rise leads one to appreciate better the complexities of nature. The concept has been popular chiefly among engineers who looked for a formula for the ground level pollution to be used without an understanding of its basis. But it is always naïve to choose the form an answer will take before an investigation is complete, and it is now widely appreciated that the formulae are valuable mainly in helping the inventor or user of them to appreciate more clearly the origin of the endless variety of situations we find. Thus formula (28) has been assumed to be an assertion that h' is proportional to U^{-3} whereas it is more profitably interpreted as saying that $h' \propto \lambda^{-2}$, because the variations in λ are much more dramatic than those in U.

Even supposing that the plume rise were a clear concept, and its value correctly known, we still have the enormous variations in ambient eddy strength to cause corresponding variations in the spreading of the now passive plume. The inhibition of diffusion by stably stratified layers does lead to one clear and simple conclusion, that whatever the nature of the effluent, if the height of the source is increased it will be trapped below fewer inversions and will be isolated above a greater number so that the ground level pollution is always decreased, whether there is a wind or not. This is discussed further in the two following chapters, and leads now to the final question about hot plumes:

PLATE 2.9. Thermals rising from a large steelworks near Pittsburgh (Penn.). The smoke is carried upwards more effectively than from the smaller sources which emit less heat

does the heat emitted have any effect when thermal convection is vigorous?

Over variegated countryside thermal convection is composed of bodies of rising warm air called thermals (see Plate 3.5), which have been exploited by birds, and glider pilots, for soaring upwards. They tend to rise from "thermal sources", which are parts of the ground where more buoyancy is communicated to the air or accumulated than elsewhere. Dry, sandy or stony surfaces, built up areas with large expanses of roofing which has a small thermal capacity, concreted airfields, and fields of dry ripe corn, are well-known thermal sources in sunshine. Lush green vegetation, wet clay soil, water surfaces, or heavily wooded areas remain cooler chiefly because of the evaporation of large amounts of water vapour into the air. The size of thermals is determined by the typical size of a thermal source, and this varies from around 400 metres square to about twenty times that area. A large modern power station emits as much heat as the input of sunshine (one calorie per cm^2 per minute) into about one square

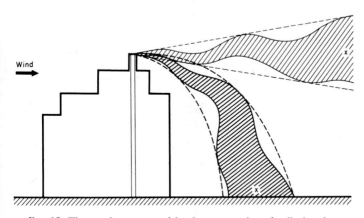

Fig. 13. The maximum ground level concentration of pollution due to a chimney on a building when the effluent is carried to the foot of the building may be computed by imagining the chimney to be isolated and computing for a free plume the pollution at the point X, the same distance from the source as the point X on the ground

kilometre on a sunny day, and therefore it can act as an almost permanent thermal source (see Plates 1.12, 2.9, 5.9), drawing in air from around it. A smaller artificial source may be smaller than the typical natural sources in the neighbourhood, and consequently it is likely to become subjected to the downdraughts in between them. To gather pollution sources together is therefore an advantage in sunny weather (see Plate 2.9).

DILUTION NEAR BUILDINGS

The main problem directly connected with buildings is the separation of the flow which takes place at their corners and the wakes which form in their lee. This is discussed more fully in Chapter 6.

If a plume is deflected down to the ground by the flow down from the top of a building on top of which there is a small

PLATE 2.10. The smoke from this lime kiln is carried by the airflow down to the ground in the immediate vicinity, because the chimneys do not protrude far enough above the building. The smoke is black because, on lighting up a kiln the temperature is too low at first to ensure complete combustion of the fuel

chimney, the order of magnitude of the concentrations to be experienced at the foot of the building can be estimated by assuming an ordinary plume, widening along a cone of angle, 6°, say, with a speed equal to the wind speed. In effect we imagine the dilution to be the same as for an isolated chimney and wrap the plume so imagined round the building and down to the ground close to it (Fig. 13). There may be residual pollution from previous whiffs when a whiff descends, and the real problem of building wakes lies not in the high maximum concentrations experienced but in the fact that often there are regions such as courtyards from which the pollution is never completely cleared even though it arrives in gusts or whiffs.

CHAPTER 3

High Level Inversions

CAUSES

INVERSIONS away from the ground have three main causes: subsidence of the upper level air which is typical of a developing anticyclone, radiation from the top of a cloud layer, and subsidence in between convection clouds.

Two of these causes may commonly operate together to produce very persistent inversions, and there are also other less common causes of inversions such as the descent of cold air in showers. Some of the many possibilities are illustrated in the examples which follow.

High level inversions set a limit to the upward dispersal of pollution. Since this is the most important single mechanism for getting rid of pollution emitted at ground level they can have a very important influence on the effects of large sources of pollution. Emissions from area sources such as conurbations are diluted more slowly under inversions (see Chapter 2) and many plumes from large single sources are also trapped below them. But since inversions often occur at only two or three times the heights of some modern large chimney stacks there is a real possibility, discussed below, that in many cases the plumes from power stations might penetrate through them.

Inversions close to the ground or in valleys are discussed in Chapter 4.

ANTICYCLONIC GLOOM

In developing or semi-permanent anticyclones the air between about 500 and 5000 metres above the earth's surface is descending slowly at a rate typically around 1000 metres per day. An individual parcel of air such as A in Fig. 14 is thus warmed 16°C, to A', and because the air is stably stratified it replaces air at A_1 that was perhaps only 5°C warmer. Some of this temperature rise is offset because the loss of heat by radiation is increased when the air becomes much warmer than that above and below it. Air

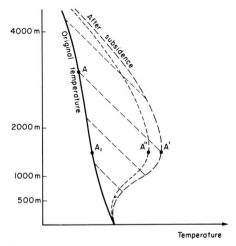

Fig. 14. Anticyclonic subsidence which is often a maximum at levels around 2000 to 3000 metres, warms the air adiabatically so that a parcel at A would change to A' in a day or two. At the same time, the additional heat loss by radiation might cool it to A'' where it is still several degrees warmer than the air at A_1 which it has replaced. The stability is decreased above A' but is markedly increased below because the subsidence decreases to zero at the earth's surface

near the surface subsides less, as does also the air at the higher levels. Consequently the stability in the upper levels decreases while it increases very markedly in the lower levels; and at the same time there is a decrease in the rate of warming of the subsiding air. In old anticyclones a steady state is reached in which the warming by subsidence is completely offset by the heat loss due to radiation.

The heat loss by radiation from the air to the layers a few thousand metres thick above and below it is small by comparison with the heat loss from clouds because the air itself only emits radiation in wavelength bands which are also absorbed by it, whereas clouds emit like black bodies and in many wavelength bands the energy is lost to space. Furthermore, when a cloud layer is established below the level of maximum subsidence further warming is completely offset by the radiative heat loss and so the cloud may be persistent for several days, producing what is appropriately called anticyclonic gloom. The following example illustrates some of the mechanisms.

A cloud layer may be formed under an inversion in one of two ways. In the first, convection from a warm surface below causes upcurrents which condense into cloud, like cumulus, but which spread out at the inversion which they are not warm enough to penetrate. This occurs over land in sunny weather and over the relatively warm sea in winter. Over land the sunshine may warm the whole convection layer sufficiently to evaporate or 'burn up', the cloud, as appears to be happening in Plate 3.2, but over the sea the surface temperature is seldom high enough for this to happen. The second way a cloud layer can be produced is by cooling at the surface: this is a much more slowly operating mechanism than the first because eddies have to be present to stir up the air while it is stably stratified on account of the cooling itself. Consequently it occurs over rough or hilly ground or when the distance travelled over the sea is very large and there is plenty of time for the cooling to penetrate upwards (see Plate 1.6). As already mentioned, once the cloud is established it tends to perpetuate itself by the cooling at its top.

In the case described in Fig. 15 cloud was produced under an anticyclonic inversion by passage of the air over England from the east during a February night. The air containing the cloud then passed westwards over the Irish sea and is shown in Plate 3.1. Below the cloud the air had become badly polluted in passing over the densely populated areas of N. England, and because only small eddies were present, the sideways spreading of this

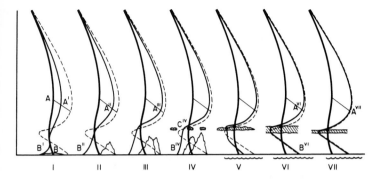

Fig. 15. Successive profiles of temperature in the development of a case of anticyclonic gloom. The initial and final profiles are shown in all diagrams, for ease of comparison.

(i) Subsidence in the upper air causes a warming shown by the descent of the air at A to A'. The surface air begins to be cooled over land from B to B'.

(ii) and (iii) The subsidence and warming aloft continue; the depth of the cooled layer increases as the height of hills and ground roughnesses increases. The air is moving across north England.

(iv) Broken cloud begins to form (C) at the top of the cooled layer.

(v) The cloud becomes an almost complete layer and cooling by radiation becomes dominant.

(vi) The cloud thickens because of convection over the sea. It is now at its thickest, and is illustrated in plate 3.1.

(vii) Subsidence continues: the cloud is lowered slightly and made thinner, the height of the base being determined by the sea temperature and the humidity. The inversion is very intense, but further subsidence could evaporate the cloud by lowering the altitude of its top to that of its base

pollution was slight as it passed over the sea. Consequently, on reaching the Isle of Man, 80 miles away, it had scarcely been dispersed. Nevertheless the cloud became dispersed during the day over England when the ground was warmer and the clearance reached the Isle of Man in the evening. The pollution was still dense at sunset (Plate 8.6) however, because the height of the inversion had not been raised.

Subsidence does not necessarily disperse a cloud layer because the radiation from the cloud top prevents it from becoming

warmed. The cloud cannot be cooled by radiation to a temperature below that produced by convection from the surface. Much
cloud and fog over the sea may have initially been formed by
cooling from below, but as soon as it is formed it is cooled further
by radiation until convection from the surface is set up. The only
way in which subsidence can cause the disappearance of a cloud
layer in this situation is by causing the inversion which places a
lid upon the convection to sink to below the condensation level.

PENETRATION OF INVERSIONS BY HOT PLUMES

It is possible to make a rough calculation concerning the
penetration of an inversion by pollution from a large hot source.
If the pollution arrives at the inversion with an excess temperature of several degrees the polluted air will rise into the stably
stratified air above the general pollution top level and will remain

PLATE 3.1. A layer of cloud beneath an anticyclonic inversion seen a
few miles west of Liverpool early in the morning. The cloud is lumpy
because feeble convection is occurring over the relatively warm sea

PLATE 3.2. Pollution accumulated below a layer of cloud at an anti-cyclonic inversion. During this summer afternoon convection from the ground was just strong enough to warm up the whole layer of air below cloud top so that the cloud became broken and very thin. This was in a hilly area but the convection was causing the flow to separate from the hillsides (see Plate 6.1, taken the same day) and the gaps in the clouds moved with the wind and did not remain over the valleys (see Plate 3.4)

there. Often there is a wind above an inversion which carries any emergent pollution away.

If the air is stably stratified down to the ground a plume emitted into it mixes with its environment as it rises and soon reaches its equilibrium level (see Plates 2.3 and 4.9). Its rise is inhibited by the mixing that occurs at the lower levels with air colder than at higher levels. In such a situation the real problem is the fumigation that occurs when the stability is destroyed by sunshine. If, on the other hand, the air is well mixed, as it usually is within a wet fog layer, and as it often is below a layer of cloud, the effluent from a large hot source will rise until it reaches the inversion at the top of the stirred layer. The eddy velocities in a layer stirred by cold convection from above are typically $\frac{1}{2}$ to 1 metre per second.

If a plume rises as a vertical cone in calm air the mean maximum upward velocity (i.e. the central velocity averaged over a long time) is given by the following equation (Rouse, Yih, and Humphreys 1952; or Scorer 1959, Table 1)

$$w_{max} = 4 \cdot 7 \; z^{-1/3} F^{1/3} \qquad (33)$$

where z is the height above the apex of the cone outlining the plume, and this is not far below the chimney top except if the chimney were unusually wide. F is the buoyancy flux and, when g is gravity, is given by

$F = g \, B \times$ (rate of emission of volume)

where $B = \triangle T/T =$ temperature excess at emission $\div$

temperature of environment.

Thus, for example, if $\triangle T = 100°C$, $T = 300°K$, and the volume emission is 1500 cubic metres per second, which is typical of a modern power station chimney, we find that $F \approx 5000 \; m^4 \; sec^{-3}$, and $4 \cdot 7 \; F^{1/3} \approx 30 \; m^{4/3} \; sec^{-1}$. Consequently, at a height of 125 metres, which might well be the height of an inversion above the chimney top, $z^{-1/3} = 5^{-1} \; m^{-1/3}$, and

$$w_{max} = 6 \text{ m sec}^{-1}$$

which is much in excess of the ambient eddy velocity produced in the convection set up by cooling a cloud top.

For a bent-over hot plume in a smooth wind of strength U, the upward velocity of the top of the plume is given (Scorer 1959) approximately by

$$w_{top} = 0 \cdot 75 z^{-1/2} U^{-1/2} F^{1/2} \qquad (34)$$

and for the case in the above example, since in a bent-over plume $w_{max} = 2 \, w_{top}$, we have

$$w_{max} \approx 100 z^{-1/2} U^{-1/2}$$

the units being metres and seconds. At a height of 125 metres, and in a wind of 4 m sec^{-1}, the maximum velocity is of the order of $3 \cdot 5$ m sec^{-1}.

These calculations show that in the cases considered the velocities in the plume are still well in excess of the ambient eddy velocities, and even at 250 metres the same is still true. The dilution *is* therefore more or less the same as in an environment with no eddies at all and we can proceed to the second stage of the calculation, which is to find the excess temperature with which the plume arrives at the inversion.

In a calm environment the maximum buoyancy in a cone-shaped plume is given by

$$B_{\max} = 11F^{2/3}g^{-1}z^{-5/3} \qquad (35)$$

where $B_{\max}$ is the maximum mean buoyancy. Thus in the case under consideration, at a height of 125 metres

$$B_{\max} = \frac{11 \times (5000)^{2/3}}{(125)^{5/3} \times 9 \cdot 81} \approx 10^{-1}$$

which means that the maximum excess temperature is still of the order of $10^{-1} \times 300°$K, which is $30°$C. The mean excess temperature in the polluted air is about a third of the maximum, and is therefore around $10°$C.

In the bent over case we have (Scorer 1959)

$$w_{\max}{}^2 \approx 1 \cdot 1 g \bar{B} z \qquad (36)$$

or

$$\bar{B} \approx 2Fg^{-1}z^{-2}U^{-1}$$

In a wind of 4 m sec^{-1} at a height of 125 m we find that $\bar{B}$, the mean buoyancy in the polluted air, is given by

$$\bar{B} \approx 0 \cdot 015$$

so that the mean excess temperature is still around $4 \cdot 5°$C in this case.

At twice the height for which the above calculation was made the effect of the wind in bending the plume over just begins to make itself felt because of the different power of z in the formula. The assumptions on which the formulae for the bent over plume are based are only valid after the plume has become properly bent over, which is the case only after the rate of rise of the plume is less than the wind speed. Thus for wind speeds less than the rate of rise we should use the formula for the vertical plume. In a wind of 5–10 m sec^{-1} the factor U^{-1} in the formula becomes important. Clearly the formula for the bent over plume cannot be applied in very light winds because this factor would have nonsensical implications.

The magnitude of the inversion above the top of a well-established cloud layer is typically between 3°C and 10°C, and so hot plumes of the kind in this example will sometimes penetrate them. An example of this is seen in Plate 3.3.

PLATE 3.3. Operation "Chimney Plumes": an experiment to see whether hot power station plumes would penetrate an inversion. The picture shows the Brunswick Wharf plume situated on the Thames, emerging through the cloud top and being carried away by the different wind at the higher level. In this experiment one plume was coloured red so that it could be identified from the air. The inversion was at a height of 800 ft and the chimney height was 300 ft

Chimneys emitting pollution with a smaller buoyancy flux at a lower altitude are at a disadvantage on several counts. Most important is the fact that the upward velocity produced by the buoyancy is reduced to less than 1 m sec^{-1} before the inversion is reached so that the dilution has proceeded further before it arrives there. Generally a smaller emission is made from a lower chimney and so its effluent has to ascend further before reaching an inversion. Inversions occur at all heights from the ground upwards, so that every increase in stack height noticeably decreases the number of inversions under which the pollution is trapped.

PLATE 3.4. The cloud is a wave cloud formed over the mountains averaging 300–600 metres high at a height of about 2000 metres in a fresh easterly airstream. The picture, taken at Harlech on the west coast of Wales, shows pollution still confined in the air below about 600 metres even though it had come from central England

THE PERMANENCE OF INVERSIONS

It is unlikely that an inversion will be destroyed by mechanical stirring from below, that is by the air moving over rough ground. Plate 3.4 shows the pollution confined below an inversion at about 600 metres even after the air had passed over the Welsh mountains in a fresh wind. When a layer of air is stirred from below the air at the bottom is warmed and the top of the stirred layer is cooled so that the strength of the inversion may be increased by the mixing.

When no cloud is present the air below an inversion may be cooled on account of the pollution present in it. Plate 3.5 shows an example of a densely polluted layer seen from above. The ground horizon is invisible because of the pollution. The flatness of the horizon indicates the wide extent of the inversion.

PLATE 3.5. A cloud free inversion seen from the air over southern England. The flatness of the horizon indicates the permanence of the inversion as a barrier to the upward diffusion of pollution

We have already remarked (Chapter 1) that cumulus clouds tend to produce a haze top at the cloud base. This is true of even quite vigorous cumulus (see Plate 3.6), especially in the morning or in the interior of large land masses. In the afternoon over England, for example, the cloud-base inversion is often slightly above the condensation level because the inflow due to sea breezes cancels out the sinking motion between the clouds which normally causes a cloud-base inversion below the condensation level. In such a case the pollution is transported by most thermals through the cloud base and dispersion is effective.

When the cloud-base inversion is below the condensation level only a few thermals form clouds, and in many cases even these rise only a very short distance above the condensation level for two reasons. First, the air above is often very dry because of subsidence in an anticyclone (see Plate 3.7), so that the cumulus are rapidly evaporated when rising into it; and secondly, there is often a sharp change of wind at such a stable level: this is seen

PLATE 3.6. Cumulus over the hills near Sheffield on a summer morning, showing the very effective mixing of the air up to cloud base but not beyond

in Plate 1.16 to be a cause of rapid evaporation of the small cumulus.

There is no prospect of artificially stirring the pollution accumulated under an inversion into a deeper layer of air because the energy required is too great. The work required to mix the air over one square kilometre confined below an inversion of 5°C at 500 metres into twice its depth is about 10^{10} m kg if the efficiency of the process were the best possible. Most stirring processes have an efficiency of about 1% and so, to reduce the pollution concentration by such means over this small area by

PLATE 3.7. An anticyclonic inversion seen from the air over central France in summer. The haze makes the ground features almost invisible. Thermals can be seen as they hump the inversion up temporarily because the haze appears denser locally when the humped haze top is viewed obliquely. Some thermals produce small cumulus clouds, but none of these grows large. Compare this with the more vigorous cumulus in Plate 1.15

only half would require energy equivalent to the combustion of 250 tons of oil fuel.

We may express this in terms of hydroelectric power: the negative buoyancy of air in a valley may be equivalent to it being 3°C (or 1%) colder than the air above. Since its density is already one thousandth the density of water, to lift the polluted air from the valley it would be necessary, if the means were 100% efficient mechanically, to let water descend from the height of the inversion and fill one hundred thousandth of the volume of the valley. In most valleys this would represent a very sizeable lake!

INVERSIONS BEHIND COLD FRONTS AND SHOWERS

From time to time passing weather of short duration produces an unusual accumulation of pollution. Occasionally a strip of high cloud may shade a strip of ground so that thermal convection occurs all around it but not over it; the result is to produce subsidence in the lower layers of air over the strip so that pollution emitted there is kept near the ground. Such strips of cloud sometimes occur behind cold fronts or ahead of warm fronts, and in some situations may be all that is left of an old front that is fast disappearing.

A cold front is the advancing boundary of a mass of cold air. It is the inclined upper surface of the cold air which is undercutting the warm air it replaces, and is a pronounced inversion through which thermal convection in the cold air does not penetrate. All wind systems in the atmosphere are set going and maintained by the release of gravitational potential energy: cold air sinks downwards and warm air ascends, so that while it advances the cold air is usually subsiding. Consequently inversions are often formed within the cold air by the same mechanism as in anticyclones. It is only when ground heated by sunshine, or warm sea, is hot enough to produce thermal convection through these inversions that the vigorous cumulus clouds characteristic of cold air masses are formed. The subsidence is usually stronger close to the front and so it is when the cold air mass first arrives that accumulations of pollution in the lowest layers of air are most likely to occur.

When a cold front passes there is often a very noticeable decrease in wind, and such a decrease is more common when there is strong subsidence taking place in the cold air. Accumulations of pollution are more likely to be obnoxious in this situation when the front passes over a town in the evening, because then more domestic (i.e. very low level) pollution is being produced and the sunshine is too feeble to de-stabilise the lowest layers of air. Plate 3.8 illustrates this.

Many big shower clouds are complete storms in themselves. They contain within them a mechanism for exchanging low level warm moist air with high level drier air. The latent heat condensed in the moist air as it rises adds to its warmth and to the

PLATE 3.8. Pollution accumulated in the lowest layers of air after a cold front has passed over south-west London in the evening

PLATE 3.9. A line of showers had passed shortly before this picture was taken and other shower clouds can be seen in the distance. A layer of pollution accumulated in the cold air as it passed over Southampton can be seen in this country area 25 miles away

intensity of the convection, and some of the rain so formed is evaporated into the drier air as it sinks to replace the warm air and thereby causes additional cooling. Many showers leave a carpet of cold air spread out on the ground behind them, and at the top of this layer there is often a very marked inversion. Unless sunshine is sufficient to warm it and cause thermal convection to penetrate the inversion, pollution is accumulated in this cold air, often more noticeably in valleys. Plate 3.9 shows an example of this.

Inversions behind cold fronts may cause mountain sides to be polluted when normally pollution is carried over the mountain tops (see Plate 4.17).

CHAPTER 4

Ground and Valley Inversions

RADIATION fog and inversions over more or less flat country were considered in Chapter 1 and in Chapter 3 we described inversions well away from the ground. In this Chapter we discuss the effects caused by valleys and sloping ground.

KATABATIC WINDS

The cooling by radiation into space of a solid surface is rapid partly because the layer of air that becomes cooled with it is shallow. Cooling by contact and molecular conductivity affects a layer only a few inches thick. In a light wind this cooling is spread into a layer many metres thick, and the radiative exchanges that take place between the ground and successive layers of air of different temperature also spread the cooling through several metres. The radiative heat transfer is mainly due to the emission and absorption by the water vapour in the air.

When the cooling is sufficient to form a thin layer of fog (i.e. cloud) on the surface the cooling is transferred to the fog top and the fog layer is thereby much more rapidly cooled.

In Plates 4.1 and 4.2 an occasion is depicted in which fog was formed on mountain slopes. The cold air on slopes drains down them as a katabatic (downslope) wind which fills valleys and hollows with pools of cold air. Plate 4.3 shows what the consequence can be by sunrise.

Katabatic winds are usually very shallow and often cause the transport of domestic pollution produced on hillside housing areas to be carried down into valleys below. Of course, when a katabatic wind is blowing there is an inversion a few metres

PLATE 4.1. Fog forming on the lower slopes of Cader Idris after sunset

PLATE 4.2. The same as 4.1 a few minutes later when the cold air had begun to drain down the slopes into the valleys

above the ground and so pollution emitted at a low level, such as domestic smoke or bonfire smoke, is trapped, and the ground level concentrations produced are rather high.

Pools of cold air in valley bottoms are one of the most common causes of the accumulation of pollution to obnoxious concentrations for long periods. In Plate 4.5 a low level inversion is seen which was kept in being for a long time by the formation of a layer of cloud above it in the morning.

HOT SOURCES IN VALLEYS

In some valleys inversions are very frequent, and the effluent from even fairly large sources may be trapped (Plate 4.6). Even if a hot plume can emerge from a smog it may still be trapped when no wet cloud is present and the air is stably stratified over a considerable depth below the pollution ceiling (see p. 59).

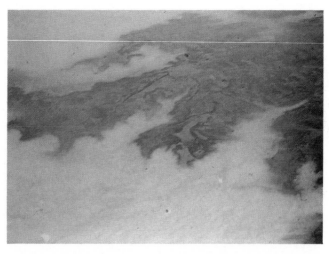

Plate 4.3. Cotswold valleys seen from the air soon after sunrise in winter. The outline of the fog represents approximately a contour of constant height on the hillside. Such fog begins to be dispersed at its edges first, for there the fog layer is shallowest and the sunshine penetrating through it to warm the ground is more intense

PLATE 4. 4. A view of Exeter from the cathedral on a sunny summer day

PLATE 4.5. Exeter on a December morning seen from the cathedral when a low level inversion caused pollution (mainly domestic) to be concentrated in the valley. The arrival of anticyclonic gloom (see Chapter 3) during the morning cut off the sunshine so that the 'smog' persisted all day

PLATE 4.6. The effluent from Hope cement works near Sheffield trapped below an inversion in a valley at sunrise in February. This phenomenon was prevented by the construction of a taller stack which was almost level with the surrounding hills (see this in Plate 2.1). The new stack can be seen under construction in this picture

The only way to ensure that the effluent is not trapped in a valley where the stable stratification is caused by hillside cooling and katabatic winds is to place the level of emission above, or at least only a very small distance below, the level of the surrounding hills (Plates 4.8, 4.9).

In the case illustrated in Plates 4.6 and 2.1 efforts were made by people interested in preserving the amenities of the country-side to keep the stack below the hills so that it could not be seen. These enthusiasts for natural beauty ought to have been more concerned about the pollution than the chimney itself, for the plume must be visible above the hills if the valley is not to be polluted. This exemplifies the principle that the forms of buildings and chimneys are not always paramount: a plume of condensed steam is as essential a part of the appearance of a cement works as the solid structures, and should be included in all sketches of it.

PLATE 4.7. The top of smog formed over Wandsworth (South-West London) during the night, beginning to be dispersed in the morning when the wind increased on the approach of a warm front. There were two or three large sources of heat (in the neighbourhood of Fulham power station) whose effluent was carried through the top of the smog and can be seen being mixed into the drier air above

PLATE 4.8. Sunrise smog in the Irwell valley in Salford seen from the high ground at Irlams o' th' Height. Agecroft power station has four large cooling towers and two chimney stacks and the effluent satisfactorily avoids being trapped within the valley because the height of emission is about level with the surrounding hills

PLATE 4.9. The plume of Ironbridge power station (see also Plates 5.12 and 2.4), in the valley of the river Severn, drifting horizontally without dispersion on reaching its equilibrium level on a calm autumn morning. The air movement was inhibited by the neighbouring hills and in this case the stratification was very stable over a great depth because of anticyclonic subsidence as well as cooling close to the ground. However the 'thermal rise' was sufficient to raise the plume above the level of most of the surrounding hills

ANABATIC WINDS

The pollution accumulated in a wide valley bottom is depicted in 4.10, where the tall stack of a chemical works protrudes above the dense layer. During the morning the heating of the ground raised this layer in the same way as it raises mist patches in meadows (Plate 4.11), but as soon as slopes become warmed anabatic (upslope) flow begins. This is made much deeper than katabatic flow by thermal convection from the surface. Only when the whole air mass within the valley is stably stratified is the heated air effectively channelled into an anabatic wind for then thermals cannot ascend far from the surface. An anabatic wind on a big mountain slope might typically be 100 metres deep (Plate 4.14). Anabatic flow stops at the snow line because the

PLATE 4.10. Low level accumulation of pollution in the very stable air in a wide valley bottom collected by katabatic winds. The disc on the stack top is designed to prevent downwash of effluent into the wake of the stack (See Plates 4.11–13)

PLATE 4.11. The pollution at a low level in a wide valley is first lifted by thermal convection in the morning. This picture was taken about 40 minutes after 4.10, and the sun was directly behind the chimney. The valley is that of the Durance in the French Alps

PLATE 4.12. The same scene as 4.13 viewed towards the sun. In the top of the picture is seen thin cirrus cloud. The much whiter layer is the sunlit haze whose top is at the snow line on the mountains. The anabatic winds carry the pollution to this level but not above. The picture shows the Montagne de Lure in the western French Alps

PLATE 4.13. The pollution in the valley containing the scene of Plate 4.10 about three hours later seen from an aircraft. The main mass of air in the valley was at approximately a uniform temperature, and was therefore stably stratified; but the pollution from the sources was carried by anabatic winds up to the snow line

PLATE 4.14. Anabatic flow visible by means of small cumulus formed in the thermals rising from the slope. Cumulus are, as can be seen, then predominantly formed over the mountain tops, and almost none are formed over valleys because the air there remains stably stratified

PLATE 4.15. Sometimes cloud is formed in anabatic winds and spreads out from the hillsides at the upper limit. In this case cloud is seen spreading out from the north (sun facing) slopes of the valley at Chamonix at the snow line. The density of haze below the inversion is much greater than above

snow reflects almost all the energy of sunshine and still radiates like a black body. It therefore remains cold.

When an anticyclonic inversion exists below the level of the mountain tops (Plate 4.16) the movement of air is inhibited and pollution accumulates in the valleys. The exact position of the

PLATE 4.16. An aerial view of the Jura Mountains looking westwards towards the sun in summer, showing the haze top, with cloud at the same level in a few places, below the main mountain tops

inversion is often determined by the snow line or by the altitude at which layer cloud is first formed.

When anabatic winds blow the pollution produced in valley bottoms is carried towards the slopes and in some localities this wind is even referred to as a 'sea breeze'.

POLLUTION ON HILLSIDES

When there is a general drift of air, pollution may still be trapped below inversions and carried into contact with the ground if there is an inversion below the highest ground. Plates 4.16 and 4.17 show examples of this.

PLATE 4.17. The plume from the power station at Wellington, N.Z., impinges on the hillsides when the flow over the top is inhibited by an inversion—in this case the inversion behind a cold front was increased in strength by nocturnal cooling

When the inversion is below the snow line the pollution may sometimes be carried higher by anabatic winds, but this is rare.

There is no practical remedy in cases like that shown in Plate 4.18, where the mountains extend up to almost 2000 metres, but in cases like 4.17 an increase in chimney height can make a very useful difference because, in a city, it significantly decreases the extent of the pollution in built up areas. The pollution does not descend at all by day, and almost certainly katabatic winds carry very little of pollution impinging on a hillside down the slope because the katabatic wind is practically independent of any general light drift of air.

DIFFERENT KINDS OF SMOG

Originally 'smog' was used to describe a natural fog made obnoxious by smoke pollution. Smogs are shown from above in Plates 4.7 and 3.3. But it is such a descriptive word that it has

PLATE 4.18. The pollution from an oil-burning cement works near Tripoli, Lebanon, engulfing orange and olive plantations on the mountain sides when it spreads out in the very stable air in an anticyclone. The mountains here are very high and therefore inversions below their tops are very common. Inversions are very common in this latitude in any case and are responsible for the well known sunny weather of the region

PLATE 4.19. Haze below an inversion in the early morning in air behind a cold front at Lower Hutt in Wellington Bay, N.Z. This area is becoming rapidly industrialised and will become liable to serious incidents if the emission of obnoxious pollution is not rigidly controlled. It is unlikely to become subject to Los Angeles-type smog because of the generally good ventilation of the area with frequent air replacement

PLATE 4.20. Los Angeles smog is unpleasant mainly because of the chemical changes which occur in hydrocarbons in the air in prolonged sunshine. The remedy is to control, and if possible eliminate, the emission into the atmosphere of incompletely combusted hydrocarbon fuels and other hydrocarbons not intended for combustion such as oil in storage, solvents used for dry cleaning, and gasoline in car fuel tanks and filling stations

been used to describe almost any nasty form of pollution. Plate 4.19 shows pollution over a town shut in by mountains, and the pollution there has been called smog. But in New Zealand the air movement is such that this pollution does not remain stagnant for very long. Over Los Angeles, however, the story is quite different because of the much greater production of gaseous pollution and the fact that the air often remains over the same area, walled in by mountains on the landward side, for many days.

PLATE 4.21. A smog over Santiago, Chile, where, as at Los Angeles, the high mountains cause the stagnation of air for several days in bright sunshine

PLATE 4.22. Santiago smog viewed from a low level. In spite of the sunshine the air is so stable that even anabatic winds carry very little of the polluted air up the hillsides

In bright sunshine many chemical changes occur which produce ozone and nitrogen peroxide from hydrocarbon vapours.
A Los Angeles smog (Plate 4.20) is not composed of smoke, and
usually, because the visibility exceeds one kilometre, is not fog.
Nevertheless it is a clearly visible haze with obnoxious properties
whose causes are now fairly well understood. There are many
other areas of the world whose development through the use of
oil will make a Los Angeles-type smog likely, if precautions are
not taken in advance to avoid it. Plates 4.21 and 4.22 show an
example of this.

CHAPTER 5

Wet and Coloured Plumes and Natural Pollution

In this chapter we are concerned with a variety of thermal, mechanical, and optical effects which influence air pollution as we see it. Understanding of the behaviour helps in the quick recognition of the composition of a chimney plume.

WHITE AND DARK PLUMES; STEAM

The wet washing of flue gases to remove almost all the SO_2 is carried on at Bankside and Battersea power stations in London. The gases are passed up a tower through which water containing alkali in solution descends and the SO_2 is dissolved. At the same time moisture is evaporated into the gases where they are hot and unsaturated at the bottom of the tower, and they may then be cooled in the upper part to a temperature below their dew point. In such a case water droplets are condensed to form a cloud. There may be some further cooling during the ascent up the chimney and this increases the density of the cloud, and it emerges dense white at a temperature perhaps only 5° to 10°C above that of the atmosphere.

Water drops are formed by splashing within the washing tower, but any of these which are carried in the gas stream are large enough to be captured in spray eliminators in the same way as they are removed from the air in a cooling tower. The cloud droplets formed by condensation in the gases are always too small to be removed in this way.

When it mixes with the atmosphere the cloud in the effluent

gas may at first become more dense (see Fig. 16), but soon the dilution is so great that evaporation of the droplets begins. The latent heat absorbed is then sufficient to cool the plume to between the ambient wet and dry-bulb temperatures, and its heat deficit may then be as great as if it had been emitted with a temperature as much as 100°C below that of the environment. In such a case the plume descends, just as it would ascend with a heat excess of the same magnitude (Plate 5.1) and often reaches the ground within as few as 5 chimney heights of the source.

The behaviour is rather variable because it depends on the amount of water evaporated in the washing chamber and the extent of the subsequent cooling before emission into the atmosphere. This, in turn, depends on the temperature ranges in the water and gases in the washing and these are determined by their

PLATE 5.1. Bankside power station plume seen from St. Paul's Cathedral at approximately the same height as the chimney top. The dense white plume first evaporates and then descends on to the river. It remains visible because of the residual SO_3 mist which, because of the smallness of its particles often looks blue. These particles also pass through the washing chamber uncaptured, and grow slightly in the atmosphere because they are hygroscopic (see Plate 5.6)

flow rates and their mechanisms of contact. The cooling after emission is equal to the latent heat of evaporation of the liquid water contained in the plume at emission, and the rate at which it takes place depends on the ambient temperature and humidity.

Distinction must be made between droplets produced by splashing in a cooling tower or flue gas-washing chamber and

PLATE 5.2. Two of the plumes from Battersea power station seen in late afternoon. The left stack is emitting unwashed, and therefore hot, flue gases, which are visible on account of their ash content. The right plume is washed and emerges as a dense water cloud which is seen here as dark because the illumination is from beyond it. (See Plate 5.5.)

The right hand plume on this occasion is only slightly buoyant and becomes bifurcated (see Plate 2.8) on being bent over. The other plume is ascending much more rapidly and is scarcely deflected from the vertical, the wind being light. The environmental turbulence is almost non-existent because of the cooling of the ground in late afternoon and the wet plume soon reaches a stable layer where it appears to become completely divided, largely because the complete evaporation of the water cloud first takes place in the central more diluted part. After this the continuing evaporation causes a rapid decrease in plume density and some descent

PLATE 5.3. The 'steam' from a train quickly evaporates and the latent heat absorbed is largely responsible for the rapid descent to the ground. (See Plate 1.2.)

Dense 'smoke' (mainly ash) from a steam engine when the fire is being drawn does not descend like the moisture-laden plume emitted when the engine is working at considerable load, when the effluent is mainly steam cooled by adiabatic expansion and has a much smaller proportion of flue gas

droplets formed by condensation due to cooling. The splashed droplets are usually of drizzle size, and if a cooling tower produces drizzle it is because the spray has not been properly eliminated. This deficiency should be cured because unnecessary water is lost from the tower by drizzle.

On the other hand droplets formed by the condensation of cloud are far too small to fall any appreciable distance through the air close to the place of emission. They travel, like smoke and pollution gases, with the air and trace out its movements. Except in the case of the rapid adiabatic expansion of the pure steam exhaust from a steam engine and of natural clouds formed by adiabatic expansion in an upcurrent in the air, condensation of droplets is caused by the mixing of unsaturated parcels of air at different temperatures. This is explained in Fig. 16.

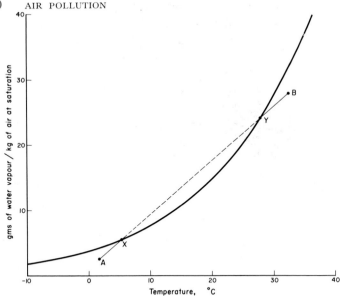

Fig. 16. The curve shows the saturation mixing ratio of water vapour to dry air in gm per kg at 1000 mb (approximately 1 atmosphere). If unsaturated parcels of air in conditions represented by the points A and B are mixed in a proportion which would produce a mixture represented by a point between X and Y on the line AB, some moisture must be condensed as cloud because the mixture is supersaturated. Thus, if a gas in the condition B emerges into an environment in condition A at first nothing is visible. But as mixing proceeds and the proportion of environmental air to effluent increases, condensation of cloud begins at Y, and the plume remains visible as a water droplet cloud until the proportion reaches ratio BX to AX, by when the evaporation is complete

If a plume emerges hot but unsaturated from a stack there may be a clear space between the stack and the visible plume. While condensation is taking place the plume gains heat and so its thermal rise is increased. This added heat is soon lost when the water cloud eventually evaporates. From the point of view of pollution at the ground, condensation of water cloud *after* emission is a good feature of a plume because it borrows the latent heat temporarily and increases the thermal rise. Condensation *before* emission by means of cooling in a stack or washing chamber is very bad because it reduces the temperature drop at the time the heat is being lost (through a stack wall, for instance) and therefore increases the total heat lost. This additional heat loss becomes real when the water cloud is evaporated, and in the case of washed plumes often causes them to become colder than the environment (Plate 5.1).

Cement works plumes show both kinds of behaviour. On a hot summer day a clear gap between the chimney top and the plume

PLATE 5.4. The cloud of condensed water vapour from a cooling tower causes no descent of the air emerging from the tower when it evaporates, because the latent heat required for its evaporation has already been gained by the gases when the vapour was condensed into cloud; and there has been no subsequent loss of heat, such as there is in washed flue gases, before emission.

Normally the air coming out of the top of a cooling tower descends very considerably in the lee of the tower because of downwash (see Chapter 4 and Plate 4.2) but the cloud is so rapidly evaporated that this descent is not visible. When the atmosphere is very damp, as on the drizzly day shown here, the cloud fragments can often be traced down to as low as the middle of the tower. The air motion is not significantly different on this day from other days with similar wind strength: if anything the descent in the lee of the tower is slightly less than usual because of the delayed evaporation

may be visible, but in winter the condensation may begin inside the stack because of the heat loss through its walls. The white plume from a steam locomotive under full power is sometimes not visible at the chimney mouth because the adiabatic expansion is still taking place or because there is a large amount of very hot flue gas in the effluent and the flue walls are also very hot, but it is more usual for a large amount of condensation to have occurred before emission.

It is easy to recognise whether or not the whiteness of a plume is due to condensed water droplets by the manner of its evaporation. As the mixing proceeds the cloud becomes fragmentary and quickly disappears in a similar way to natural small cumulus clouds. By contrast smoke becomes gradually invisible as the gas containing it is diluted.

PLATE 5.5. A cement works plume near Cambridge seen in illumination from behind on a day with variable and dense clouds above. Although it looks dark because of the lighting it consists mainly of water cloud droplets, and becomes less dark when they evaporate (see Plate 5.2). From the other side it looked lighter than the background because the sky behind it was dark (see Plate 2.1); and in a dense water cloud, in which the visibility is at most a few feet, most of the incident light is scattered (reflected diffusely) and not transmitted

PLATE 5.6. Two plumes at Clarence Dock power station, Liverpool. The left plume is from a PF boiler and is visible because of the fly ash which passes through the electrostatic precipitators. The right plume is visible because of the SO_3 mist it contains and is from a chain grate boiler: it is bifurcated, which shows that some early thermal rise is being achieved. The SO_3 particles do not evaporate because they are hygroscopic

A rather different behaviour is shown by a plume of SO_3 particles. There is always some SO_3 in the flue gases from a furnace burning sulphur containing fuel. The design of the boiler affects the appearance which has very little to do with boiler efficiency. In Plate 5.6 we see two plumes from a power station in Liverpool, and only one emits a dense white plume. This is from a chain grate boiler and the other is using pulverised coal. In this case there may be effects produced by the fact that the coal had previously been soaked in sea water, but the behaviour of the white plume is nevertheless characteristic of one made visible by an SO_3 mist. Similar but smaller plumes are common at oil refineries, and sometimes these are almost clear at the stack mouth which shows that condensation is occurring after emission.

The whiteness of an SO_3-laden plume is determined as much by the size of the particles as by the amount of SO_3 present. The effect of boiler and flue design is not primarily to alter the amount of SO_3 present but rather to determine the size of the particles into which it condenses. The plume in Plate 5.6 seems to have achieved about the maximum whiteness possible as a result of the rate at which the condensation of SO_3 happens to occur. From some flue gases the condensation is largely on to the flue or stack walls and it depends probably on the occurrence of dilution within the flue, cooling at the walls, or standing eddies at corners or joins within the flue, as well as on the furnace.

PLATE 5.7. A black 'smoke' plume which takes its colour mainly from the black carbon particles it contains. This plume is typical of a coal furnace which is intermittently stoked or of any furnace still cold after being lit. The particles are generally pure carbon formed by the reduction of CO or CO_2 in the presence of unburnt vapours above the combustion bed or flame and are not obnoxious like domestic smoke which is composed of tarry components of the fuel condensed from vapour without combustion. The introduction of secondary air above the flame can prevent this.

The plume is fragmented by the eddies formed in the flow of air over buildings

PLATE 5.8. An oil fire near Los Angeles producing very black smoke, but even this appears 'white' in suitable illumination because of the great density of small particles. This plume levels out at the inversion which is responsible for the Los Angeles smog (see Plate 4.20)

SO_3 plumes are blue when the particles are very small (see Plate 5.14), achieve a maximum whiteness at some larger size, and then rapidly become virtually invisible when the same amount of SO_3 is condensed into still fewer larger particles. As dilution proceeds the disappearance is often slower than would be the case with smoke because the particles are hygroscopic and grow in the atmosphere as they become spaced out.

There are no significant latent heat effects in SO_3 plumes because of the small amount of material involved.

The densest SO_3 plumes appear dark, sometimes brown, when seen in illumination from behind, but the more tenuous ones tend to be white in a greater variety of lighting patterns. In general plumes tend to take on the colour of the particles which make them visible when there is single scattering. But when a ray of light impinges on several particles in its passage in the plume before emerging it appears light in illumination from behind the observer and dark in light from beyond it (Plate 5.8).

There is a great variety of phenomena which stem from the behaviour of artifically introduced water droplet clouds in the atmosphere. In Plate 1.12 the cloud formed over a power station was seen to form at a lower altitude than natural clouds because of the water vapour added by the cooling towers. On that occasion there was very little condensation of cloud close to the cooling towers because of the dryness of the surrounding atmosphere. In Plate 5.9 the same power station is producing a cloud which is continuous from the cooling tower up to the altitude of natural clouds.

There are two main causes of this behaviour: first the humidity of the atmosphere was rather high, such as it might be when the ground is wet after rain and scud at a level below the main cloud base is formed. The second cause is the absence of wind: for at Hams Hall there is a group of 3 power stations and the production of heat there causes an inflow of air from all sides and an ascent of the hot air with very little mixing. (For a fuller discussion of these

PLATE 5.9. Cloud formed by the cooling towers of Hams Hall power station (see also Plate 1.12) in a situation in which it does not evaporate before reaching the altitude of natural cumulus clouds

PLATE 5.10. Contrails formed from the water vapour in aircraft exhaust seen in the evening above a layer of pollution accumulated in the stable air close to the ground. A power station plume rises higher because of its large heat content

mechanisms the reader is referred to the treatment of scud in Scorer and Wexler's *Cloud Studies in Colour*, and converging hot plumes in *Natural Aerodynamics*, page 213).

In sufficiently cold air the exhaust of an aeroplane forms a contrail (condensation trail) a few hundred feet behind it by the mechanism described in Fig. 16. But many trails become frozen because of the low temperature (which is usually necessary for their formation in the first place), and since ice has a lower vapour pressure they do not evaporate or do so very slowly (Plate 5.10). (For fuller discussion of the behaviour of contrails see Scorer and Wexler's *Colour Encyclopaedia of Clouds*.)

The steaming of hot wet surfaces illustrates some of the same mechanisms. Plate 5.11 shows a lake in the Jura Mountains steaming like a bowl of soup. The previous evening a cold front had passed and the water was warm at this time of year (August): katabatic winds produced a pool of cold air over the lake from which the steam shows thermals to be rising. The same kind of condensation often occurs over rivers and reservoirs whose water is circulated through heat producing industrial processes. Plate 5.12 shows an example of this. It also occurs on wet roads warmed by the sunshine after a shower.

The production of cloud by the mixing of two unsaturated masses of air of different temperature is also important in the production of natural fog. If the air were calm dew would be deposited on the ground and no fog would be formed on a clear night. But when the air is stirred up by the wind flowing over ground unevennesses the shallow layer very close to the ground is mixed with warmer air above and condensation occurs within the air rather than on to the cold surface. The same mechanism is responsible for the formation of water droplet cloud within flue gases, at a cement works for example, rather than entirely on the inside surface of a chimney stack which is cooled from the outside by the atmosphere.

PLATE 5.11. The water of Lac Genin steaming because its temperature was several degrees above that of the air above it. The air close to the lake becomes saturated at the temperature of the lake, and because of its buoyancy rises and mixes with the much colder air above (see Fig. 16). Thermals can be seen in the form of the top of the rising cloud

PLATE 5.12. The river Severn at Ironbridge on the same morning as that depicted in Plate 4.10. The valley was filled with cold air after a clear night at the end of September and the river was warmed by the power station. This shows, better than Plate 5.11, the characteristic appearance of the steam over a hot, wet surface. The rising parcels of air produce a shearing motion which draws out the cloud into vertical sheets and filaments

COLOURED PLUMES OF SMOKE, MIST, OR DUST

Coloured plumes may take their colour from the particles of smoke that make them visible. Red oxides of iron in the form of particles of around micron size are the cause of the red smoke often seen at steelworks (Plate 5.14) and coloured pyrotechnics for use by day sometimes produce coloured smokes.

A smoke is a cloud of solid particles which have condensed in the gas which carries them, and they are of such small size that their fall speed relative to the air is negligible compared with the air motion and they travel as a gas mixed into the air. Domestic smoke consists largely of volatile components of the fuel condensed in the flue without being burnt. Smoke can be burnt by the maintenance of a sufficiently high temperature in the presence of sufficient oxygen. At high enough temperatures CO is the only

PLATE 5.13. The red plume from a steelworks at Scunthorpe. The colour is that of the particles which are oxides of iron. These very red plumes are produced by oxygen lancing, in which oxygen is bubbled through molten steel

PLATE 5.14. The blue colour of some domestic and bonfire smoke is caused by the smallness of the particles. Since this is the colour of the scattered light it is most evident when the smoke is seen against a dark background

combustible material emitted and even this can be burnt if sufficient air is supplied. Too much secondary air, i.e. air introduced above the fuel bed, is wasteful of heat because the surplus air is warmed and then passed up the flue.

Domestic fires and bonfires usually produce smoke because there is far too much air and the volatile parts of the fuel are not all burnt. The white smoke of bonfires often consists of condensed steam from the wet material in the fire: sometimes it can be seen to evaporate as mixing with the air takes place, but at others there are so many hygroscopic particles produced that the water vapour is condensed on droplets which do not evaporate. The colour of blue smoke (Plate 5.14) is due to the smallness of the particles (of the order of $0 \cdot 1\mu$) and is the colour of the scattered light. The colour is therefore only seen clearly when the background is dark.

Sulphuric acid mist may produce a blue colour when the particles are very small. In spite of being very hygroscopic these particles are not captured in the flue gas washing process (see page 86) and often the plumes of Battersea and Bankside power stations (Plates 5.1 and 5.2) show a blue colour because of this when the water cloud has evaporated. A cloud of condensed particles is called a mist when the particles are liquid, and are easily produced by vaporising a liquid which has a very low vapour pressure at atmospheric temperatures, such as oil (which has been used for making 'smoke' screens). Sulphuric acid mist is not always blue and as produced in power stations is more usually white. Such a dense plume as that in Plate 5.6 has a rather small range of drop sizes and when the sunshine passes through it it often acquires a magenta rather than a brown tinge. Domestic smoke, on the other hand, appears brown when seen against a bright sky.

The reddish brownish colour of some power station plumes is the natural colour of the small fly ash particles which pass through the electrostatic precipitators. These particles are dust which may have been fused into hollow spheres, and since they were not condensed out of vapour are not correctly called a mist or a smoke.

The blueness of air seen in valleys and the apparent blue colour of distant dark mountains is the colour of the light scattered by the intervening dust particles.

NATURAL AIR POLLUTION

There are many kinds of naturally produced particles other than natural water clouds which obscure the air. In middle latitudes perhaps the most important is salt haze from the sea (Plate 5.15). It is produced mainly from the fragmentation of the thin film of water which surrounds froth on the sea or by the throwing up of very tiny droplets when air bubbles, trapped in the water by breaking waves, break the surface. The sudden decrease of surface area releases energy which produces very small particles whose water is largely evaporated before they fall to the sea surface. Salt haze is not produced from the ordinary spray blown off wave tops nor generated by splashing. Although

PLATE 5.15. Salt haze at Saltburn (Yorkshire) in a fresh onshore wind. It is the airborne salt which is mainly responsible for the lop-sided growth of coastal trees. Young shoots on the seaward side are stunted by the salt captured on them

PLATE 5.16. A dust devil in Uganda carrying large dust particles upwards in a strong vortex. These particles generally fall out rather soon, and dust devils usually last from a few seconds up to half a minute or so

the effects of such spray are very obvious when it falls on solid objects it is not a significant direct source of salt haze because the particles are too large to remain airborne long enough for them to lose their excess water.

Exceptionally strong onshore winds in summer can do serious damage to growing vegetation for a few miles inland on account of the salt particles captured by the tender growth.

Dust can be made airborne in three distinct ways. Dust devils (Plate 5.16) are vertical convection currents which occur more readily over a surface of loose dust because it has a very low thermal conductivity and therefore achieves a much higher temperature than most other surfaces in sunshine.

PLATE 5.17. A Harmattan-type dust haze seen at Karachi, hundreds of miles from the source of the mica-like dust from which it was formed. The sky takes on a yellow hue and the sun becomes invisible before it sets

PLATE 5.18. A not very intense blizzard in Greenland. The worst blizzards are those which raise the loosely packed snow in a valley rather suddenly. Sometimes in a valley a pool of cold air formed by katabatic winds is disturbed by a gale blowing above it and the visibility is reduced to a few feet in a very short time

Cold fronts travelling across a sandy or dusty surface some-times produce a sudden increase in wind strength and, perhaps more important, a sudden increase in the intensity of thermal convection when the cold air arrives over very hot sand. The wall of dust so produced is called a haboob (Plate 5.19) and is probably most common and certainly best known in the Sudan.

Dust is, of course, raised by a strong wind anywhere, but if there is a source of very small dust particles a haze of very long duration is produced and may be experienced many hundreds of miles from the place where the dust was raised. The most notorious is the Harmattan, a haze carried into northern Nigeria by winds from the Sahara (see Plate 5.17).

A blizzard (Plate 5.18) is another example of 'pollution' raised by the wind. It is a storm of drifting snow which is not falling from the sky at the time. Like sandstorms they are insidious because they can be raised out of clear blue skies.

PLATE 5.19. A haboob advancing across hot desert. The dust is raised by the sudden increase in turbulence on the arrival of cold air. These storms are usually accompanied by rain, but the rain is normally much less widespread than the haboob it produces

Clouds of living organisms carried by the wind have been the subject of mention in early literature because they constitute a plague upon humanity. Perhaps the best known is the desert locust whose great swarms are formed, maintained, and transported by the motion of the air in which they habitually fly. The recent control of this pest has been achieved largely through an understanding of how the locusts unconsciously exploit the air movement to the advantage of their species.

PLATE 5.20. A swarm of desert locusts seen from the air. These swarms containing perhaps 3000 million locusts with a total weight of 1000 tons, which eat vegetation equal in weight to their own each day, are carried by the wind in such a way that they arrive frequently in desert areas recently rained upon and therefore suitable for the hatching of their eggs

CHAPTER 6

Separation

CAUSES OF SEPARATION

SEPARATION is the name given to flow which leaves a solid surface. There is usually flow along the surface towards a line of separation from both sides and at the line of separation the velocity is zero unless the surface has a sharp edge there, called a salient edge.

When the flow of a fluid is diverging so that the particles of fluid are being decelerated in their trajectories, they are usually travelling towards higher pressure (according to Bernoulli's equation). If the flow is very viscous and slow the viscous forces may be enough to produce the deceleration, and this is true of the slow moving fluid close to the boundary, but further away where the effects of viscosity are negligible and the velocity greater, the fluid undergoes a smaller deceleration under the same pressure gradient. In Fig. 17 is shown the pattern of motion past a spherical obstacle when there are no viscous effects. Around the widest section the velocity is a maximum, and therefore the pressure a minimum; consequently as the fluid passes round towards the rear stagnation point, which is a point of separation, it becomes decelerated and must therefore move towards higher pressure. In Fig. 18 we see the consequences of this pressure rise when the fluid close to the surface is decelerated also by viscosity: it comes to a standstill on the circle SS, beyond which the pressure forces cause a reversal of the flow. This circle therefore becomes a line of zero velocity with the fluid approaching from both sides and moving away from the surface. The flow is thus separated from the body.

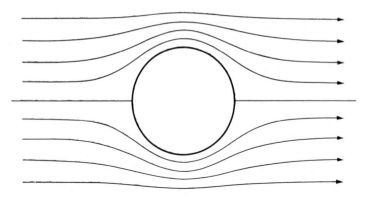

Fig. 17. The pattern of flow round a spherical obstacle when there is no viscosity. The flow separates at the rear point of the body

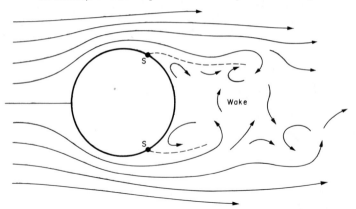

Fig. 18. The effect of viscous forces is to bring the flow to a standstill at S, where separation takes place

The chief consequence of separation on a body is the formation of a wake, in which the motion is turbulent, that is to say it contains unsteady motion with rotating bodies of fluid called eddies.

If a hill or obstacle has a sharp edge anywhere near the point at which separation would occur if the shape were rounded the, separation becomes fixed at this edge, called a salient edge (see

PLATE 6.1. Separation of the flow from a hillside somewhat to the lee of the crest of the hill. The point of separation moved up and down the curved part of the hill over a distance of about 100 yards (see Fig. 21). The wind was from the right over the hill crest, but up the hillside from the left. The smoke passed almost horizontally across the valley

Plate 6.5). The flow may rejoin the surface again further downstream (see Figs. 19, 20, 21).

In a wake region the pressure is lower than on the front of the obstacle so that the wind produces a force on the obstacle in its direction. In some cases the eddies formed in the wake are shed, and travel away downstream, producing a gustiness of the wind there, and when they are released from the obstacle the line of separation moves about over the surface and causes fluctuations in the force on the obstacle. The humming of telephone wires in the wind is caused by the almost periodic shedding of eddies and the vibrations caused are transmitted to the pole which acts as a sounding board for the wires. However, the forces on the obstacle when eddies are shed may be less than the forces on another obstacle downwind produced by the passage of the eddy. Consequently it is dangerous to place a chimney close to another because the wake of one might cause excessive vibrations in the other when eddies are shed (see Plate 8.6).

DOWNWASH

The flow behind a tall chimney is rather like that shown in Fig. 18. At the top, effluent may be entrained into this wake because it is a region of lower pressure than the environment. In that case (see Plate 6.2) smoke may be seen in the wake several metres from the top. This not only causes a blackening of the chimney, but also has the same effect as lowering the height of it as far as ground level pollution is concerned. Devices have been employed to prevent downwash. For example flat horizontal discs have been built at the top with a diameter equal to about three chimney diameters (see Plates 4.10 and 4.11). This has the

PLATE 6.2. Downwash of effluent in the wake of a chimney caused by too low an efflux velocity

effect of not allowing the effluent to enter the wake until at least one diameter downstream, and at that point it will be carried away by detached eddies and not entrained down the lee of the chimney in attached ones. The problem of downwash arises mainly when a chimney serves a boiler on low load, for then the efflux velocity is much reduced. This problem is solved by the construction of multiple stacks (see Plate 8.5).

Downwash is generally avoided by having a sufficiently large efflux velocity, and experience shows that this needs to exceed the wind speed. Some devices employed to avoid downwash on ships funnels are not usually suitable for land use because they depend on the wind direction being from the bow quarter.

SEPARATION ON CLIFFS AND MOUNTAINS

At the foot of a cliff facing the wind the airflow diverges so that an eddy may be formed there, with the flow rejoining on the cliff face (Fig. 19). The cliff top often behaves as a salient edge with an eddy extending varying distances downwind according to the wind strength, gustiness, and stability.

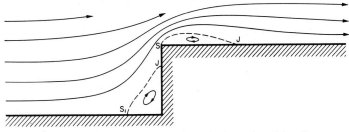

Fig. 19. A 'bolster' eddy sometimes forms in the region of slow flow at the foot of a cliff, and others may form at salient edges at the top. S is the line of separation, and J is the line of joining which usually moves about irregularly.

Separation may occur at many places on a hill of complicated shape so that eddies, shed and unshed, of many different sizes may occur in its lee. This means that the effluent of a chimney situated there is carried very quickly to the ground if the chimney top is in the wake (Plate 6.3).

PLATE 6.3. The plume from a cement works near Vancouver carried immediately to the surface in the lee of a wooded hill

The rounded surface of a small hill may equally cause separation (Plate 6.4) and a large eddy may be found in the lee of a large cliff. In Fig. 20 a situation which arose at Hope cement works (Plate 4.6) is shown diagrammatically: the line at which the flow rejoined was close to the chimney so that the effluent descended to the surface. This no longer occurred when the taller chimney (Plate 2.1) was built.

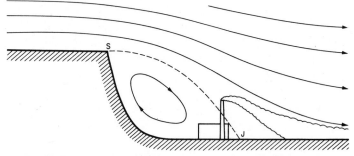

FIG. 20. When the line of joining at the rear of a cliff eddy was close to a chimney the effluent was brought down to the ground

PLATE 6.4. The flow at a separation line on the top of a small rounded
hill revealed by placing a smoke generator a small distance downwind
of the line

When the flow separates from a mountain surface the question
arises where the flow will rejoin. In the case illustrated in Plate 6.1
the smoke released close to the point of separation could be seen
travelling horizontally for a mile or so across the valley which
was between two and three miles wide. The flow was rather like
that shown in Fig. 21 in which the sunshine on the lee slope is
shown as producing an anabatic wind which makes separation
near the top of the slope inevitable.

When a katabatic wind blows down the lee slope there is a
shallow layer of cold air on the surface which cannot flow away
from it. Consequently a katabatic wind prevents separation and
induces the airflow to follow the shape of the ground.

In Fig. 22 we envisage a case in which the flow follows the
ground shape except for a small closed eddy induced by the
presence of a salient edge. The motion of the upper air will also
descend over the valley, and this will produce a permanent hole
in any thin layer of low cloud which may be present. On the
occasion of Plate 6.1 such a cloud layer existed but the gaps

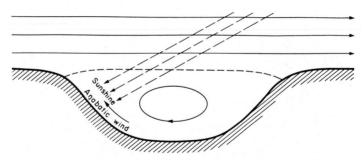

FIG. 21. A valley filled by an eddy which was in part produced by an anabatic wind on the upwind slope

between the clouds moved along with the wind and did not remain stationary relative to the ground. The scene is shown in Plate 3.2, and there was enough sunshine to produce the anabatic wind.

SEPARATION ON BUILDINGS

The sharp corners of buildings can behave as salient edges. In Fig. 23 the typical separation pattern on a flat-topped building is shown. The pollution from a chimney from within the wake region can be carried to anywhere on the building surface behind the salient edge. Thus if dense smoke were produced by chimney C_1, the outline of it would be the dashed line starting from the

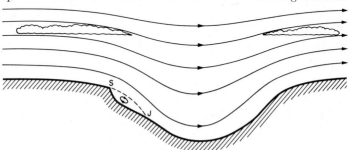

FIG. 22. When an eddy does not fill a valley the wave motion often produces stationary cloud gaps. Katabatic winds inhibit separation and may cause rapid rejoining after a salient edge

separation line S. Pollution created at B might, in strong rather steady winds be detectable at A because it would be carried by the eddies into the whole wake region. But when the wind itself is unsteady the edge of the wake moves about, and will sometimes descend on to the roof, scouring pollution away by means of gusts of fresh air from above. In Plate 6.5 the smooth wind case is illustrated: smoke from a generator placed behind a wall about two feet thick is seen penetrating forward as far as the forward edge of the wall, corresponding to the point S in Fig. 23.

In order to avoid polluting the lee side of a building with effluent from chimneys on the roof, the stack top must be outside

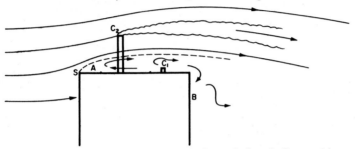

Fig. 23. Separation usually occurs at the forward edge of a flat-topped building and smoke should be released outside the wake (C_2), otherwise it might penetrate to all parts of the wake

the wake region, as the chimney top C_2 is in the diagram. There is much cosy domestic architecture, typified by Plate 8.3, in which even chimney pots have been abandoned for appearance's sake. Consequently the smoke emitted often enters the wake of the chimney and also the wake region formed by separation at the apex of the roof. In Fig. 24 the smoke is shown moving in whiffs everywhere in the lee of the house. The house itself will produce a gusty wind for the next house downwind which will make certain that, even if in a steady wind the smoke would just escape the wake region, it will from time to time penetrate all over its lee side, often entering its windows!

The occasion shown in Plate 6.6 was seen by chance from a train. Dense white smoke was emerging from a house chimney,

PLATE 6.5. A smoke generator was placed behind a wall about 5 feet high and two feet thick standing close to the sea on a rocky coast. The wind was from the right in the picture, and the smoke is seen to penetrate in the wake region right up to the forward top edge of the wall

PLATE 6.6. The track of domestic chimney effluent made visible by a chance emission of dense white smoke caused by a chimney fire. It frequently descends in whiffs with very little dilution into the street. Invisible emissions follow similar tracks

probably because of a fire in the soot accumulated in the flue. The dense smoke is a tracer of the air flow and shows that from time to time it travels upwards and then down into the street. The importance of this, (see also Plate 5.14) is that even when there is no visible emission the flue gases follow a similar path. This is one of the reasons why domestic emissions are so highly objectionable in comparison with emissions from tall chimneys: they are carried with rather little dilution into the street and into neighbours' windows so that they are breathed and come into contact with paintwork, fabrics, and garden flowers at rather high concentrations.

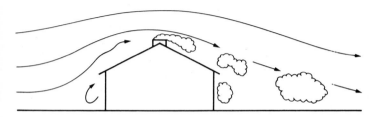

FIG. 24. With no chimney pot and a squat stack effluent usually enters the wake of a building, which means that it can enter the windows

The problem of short chimneys on buildings has scarcely entered the consciousness of architects. There is very little evidence in houses designed in the twentieth century that any serious attempt has been made to construct domestic chimneys which satisfactorily perform their function of removing the effluent so that it is not smelled in the garden and street. Generally chimneys are hidden, or placed below the apex of the roof where their smells are certain to enter a wake region. The fashion of not having tall chimney pots encourages downwash and many other evils such as cold inflow (see below). If chimneys need to be short because architects cannot make them aesthetically acceptable otherwise, there are many devices which can be employed to prevent separation at salient edges. One of these is shown in Fig. 25: it is a kind of Handley-Page slot, and has the effect of producing a shallow jet of air across the rooftop. Further improvements would be made in the flow if the corner of the building were rounded because a small region of separation might occur there even in the situation illustrated.

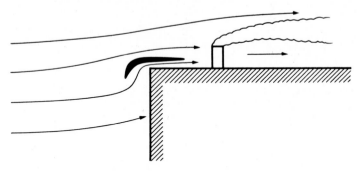

Fig. 25. An aerofoil placed at the top corner of a building to inhibit separation there might make tall chimneys unnecessary

In Fig. 26 the plan of the wake region of a tall block is shown. In a smooth wind the wake would occupy the region shown, with separation occurring at two opposite edges.

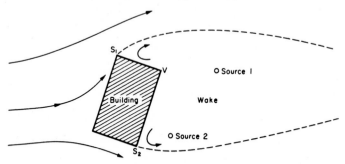

Fig. 26. The rear surface of a building can be polluted by sources anywhere in its wake. If the wind is gusty fresh air penetrates the wake more often

Pollution from sources in positions shown could be found anywhere in the wake, and obnoxious smells could be detected anywhere over the two near faces of the building.

If the wind were gusty, as on a fresh showery day, the wind direction would change frequently so that it would carry fresh air across the face S_1V from time to time. At other times gusts could scour pollution from the face S_2V, in either case the edge V would be a salient edge temporarily. The only pollution which could be detected all over both faces would be that emitted close to V, because there would not be time for the pollution from source 1 to be carried forward on to the building by eddies between the scouring gusts, and pollution from source 2 would not be detected on face S_1V except in a very smooth wind.

In many wind tunnel tests of model buildings it is important to give a correct gustiness to the wind. Wind tunnel technicians prefer to create as smooth a wind as possible, but even a naturally smooth wind is made unsteady by other buildings upwind of the one under test. The model tests usually indicate much worse effects due to the presence of buildings than are experienced in practice not because of any fundamental difficulty of modelling but because the wind eddies are not modelled at all. Very small ambient eddies do not matter. Very large ones only make slow variations in the wind so that the flow pattern is almost the same as if the flow were steady. Eddies ranging in size from about one tenth to about ten times the building size make a great deal of difference to the forward penetration of pollution in the wake region, and since these are common in the atmosphere they cannot be ignored without serious error.

COLD INFLOW INTO CHIMNEYS

If a chimney were filled with hot gas with no upward velocity the cold air from outside would flow in at the top and gradually replace it. On the other hand if the exit velocity is sufficient no such inflow takes place. There is therefore a critical efflux velocity below which inflow at the top begins. The inflow is caused by the density difference between the inside and outside and so the critical velocity is higher the warmer the effluent gas.

Normally the velocity of the flue gas is greatest near the middle of the chimney and falls to zero at the wall. It depends very much on chimney design, how thick the slow moving layer close to the wall is, and the critical velocity therefore depends on the velocity profile as well as on the density difference.

Plate 6.7 shows the cold inflow into a model chimney just beginning. It is made visible by the introduction of smoke at the rim, and it is always observed that the inflow occurs close to the wall.

The inflow does not take place always at the same point of the rim but moves round from one point to another when there is no cross wind. In the presence of a wind the inflow generally occurs at the sides and never at the upwind or downwind points of the rim.

PLATE 6.7. A model chimney consisting of a glass tube insulated by an asbestos blanket from which hot air is emerging. As the efflux velocity is reduced below a critical value, inflow of exterior cold air begins at the walls where the upward velocity is least

When the velocity is somewhat below the critical value the cold air penetrates some distance into the chimney. Since it mixes with the hot gases the density difference is gradually reduced until it cannot penetrate any further. Plate 6.8 shows this in a model experiment. The importance of this penetration is greater when the flue gases contain substances which change in character with temperature. Thus if the interior temperature falls

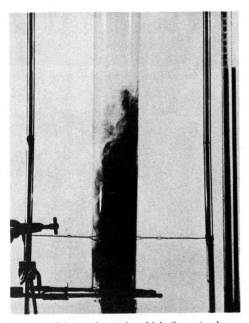

PLATE 6.8. A model experiment in which (heavy) salt water flowed out of the bottom of a glass tube immersed in a tank of fresh water. Dye was introduced at the bottom of the chimney so that the outflow and any inflowing water became visible. When the velocity of efflux was sufficiently reduced 'cold inflow', i.e. buoyant fresh water, penetrated up the tube in surges which reached to about the same point before being carried down again in the general flow.

Unlike a chimney in which the cooling of the stack wall by cold inflow would tend to locate the inflow at the same point all the time, there was no analogous effect in this experiment and the point of inflow moved erratically around the rim of the tube

below the sulphuric acid dewpoint a mist may be formed. This has little effect in itself but persistent cold inflow can cause a cooling of the stack wall, and this tends to cause the inflow to continue at the same spot and maintain the low wall temperature. Condensation occurs on the wall, which becomes wet with acid: carbon particles then become attached and are subsequently removed by the gas motion and emitted from the stack as acid soaked smuts which are very corrosive.

To prevent cold inflow it is important to increase the speed of the effluent close to the wall. This can be done in a variety of ways. If a nozzle is placed on the chimney the whole flow is speeded up and it is therefore a wasteful method because generally a substantially increased forced draught is then required.

PLATE 6.9. The diverging top of this chimney on a London block of flats causes such a low efflux velocity that downwash and cold inflow are made inevitable. Within 6 weeks of the flats being occupied the outside of the chimney had become badly blackened!

CHIMNEY TOP DESIGN

For the reasons made clear above it is desirable that a chimney top should be cleanly shaped. Elaborate decoration or increase in exterior size increases the likelihood of downwash into the chimney wake. The efflux velocity should not be unduly increased, and should be enough only to prevent downwash into the wake and cold inflow. Since the simplest design is the best it may be asked why any remarks on this topic are needed. The answer lies in what we see on buildings.

Chimney pots have some of the required features and should not be omitted from brick stacks on houses because these are thick walled and encourage downwash. Sometimes lids are placed on brick stacks, and this makes downwash inevitable. When cold inflow occurs it causes acid condensation which tends to destroy the stack structure and so it must be prevented on domestic chimneys, when new heating equipment is installed with a lower efflux velocity than before, by the provision of a suitable chimney pot. The prize for idiocy goes to the chimney illustrated in Plate 6.9. In attempting to achieve a pleasing appearance the designer has succeeded in making downwash into the wake of the building certain, which means that the upper flats of the block are frequently polluted. Cold inflow is also made inevitable.

CHAPTER 7

Some Effects of Air Pollution

ALTHOUGH we can all agree that air pollution is undesirable, and that most of its effects on health and damage to materials and amenities are well known, the pattern of our civilisation makes it unavoidable. The purpose of this chapter, is not to present the terribly familiar catalogue of misery and squalor but to draw attention to mechanisms which illustrate that air pollution produces its effects in many complicated ways. Some are understood, others at present are not: the effects and the remedies are costly, and therefore any advance in understanding will make the correction easier.

SMELLS

Brickmaking and oil refining are two industries which make nasty smells. Some of these smells do not cause any direct damage, but smells are undesirable as such and are as much a nuisance as noise, dirtiness, or corrosion. They affect people's frame of mind.

Unfortunately no instruments have been devised for measuring the causes of smells quantitatively, and evidence is very difficult to obtain. Smells are very subjective, partly because a person's attitude to the cause of a smell affects his view of the smell itself. Employees in an industry associate smells with their source of income and soon ignore them; farmyard smells are tolerable in their right context only, but any psychological accommodation to an inevitable smell is a hardening of sensitivity and destructive of at least some aesthetic subtleties, and we have no right to impose smells on others on the grounds that if they had the right mental attitude to them, they would not object.

There are also physiological mechanisms at work. The smelling organ becomes insensitive to smells to which it is continuously subjected. Thus oil refinery employees become physically incapable of detecting some smells however much they may correct any psychological adjustment that would be occurring anyway. For this reason it is useless to leave the detection of smells in the hands of people who work long hours near the source.

When a nose has been rendered more or less insensitive to a smell by continued strong doses of the substance causing it, it becomes completely insensitive to the same chemical in much smaller concentrations. But some smells become different when much diluted, and may even become more objectionable so that only at low concentrations do they give rise to complaints. It might therefore be falsely argued in defence of a smell that it was not found objectionable at higher concentrations nearer to the source. The complete causes of these low concentration effects are not fully understood but one contributing factor is certainly the complete absence of it in between whiffs.

The wind varies with height and there can be pockets of slow-moving air among buildings, within trees, bushes, and tall crops, which retain traces of a smell for long periods after a dense whiff has passed by. Close to a source of a smell, there may be no moments at all during which the smelly substance is completely absent from the air we breathe, so that our noses become insensitive to it. At greater distances, there are longer periods between whiffs and all the pockets of slow moving air become completely scoured out so that we breathe fresh air, or at least air free from the smell in question, and when another whiff arrives it is very noticeable, even though the concentration is much less than the lowest achieved closer to the source where it is not detectable.

The whiffs seem much more objectionable too, when they are less frequent. Noises at unpredictable widely spaced intervals are more disturbing than much louder but more or less continuous or regular ones; and so it is with smells. The important point here is that smells are not diluted into the atmosphere in a simple manner. Of course, the concentration of pollutant in bodies of air containing it gradually decreases but they are as sharp edged as ever.

The outline of a smoke plume or cloud is usually sharp to begin with and only seems to become more diffuse because of the decreasing difference between inside and outside. Nevertheless, we should still speak of whiffs because only after many hours in the air is a pollutant so diffused that it is no longer possible to define the boundary of the air containing it. A few seconds after emission, whiffs of smoke or smell are clear cut, and a few minutes later, the sudden arrival of a smelly whiff three miles downwind can be equally sudden and very objectionable. We may refer to Fig. 6 (page 32) to see that the further we are from a source, the less frequent is the passage of the plume over us and the lower is the concentration when it does arrive, but in the geometrical sense, the arrival is similar. If an instrument were to be devised to detect smells at large distances, it would be important not to take a longer term average further from the source as would seem appropriate for other reasons, but to measure the rapidity with which the maximum concentrations in the whiffs arrive. The whiffs probably seem to pass over more quickly than they actually do because the nose is most sensitive to rapid increases in concentration and soon fails to detect a uniform level.

HEALTH AND OTHER AMENITIES: SOME MISCONCEPTIONS

Contentment is an important contributor to good health, and therefore, any disturbance by smells, noise, dirtiness, or even appearances, impairs health. We shall not discuss any medical aspects of the problem here but merely mention some causes of error and misguided effort.

Undoubtedly, the direct harmful effects of smoking on the health of the general public are far greater than any due to air pollution. Only in special circumstances does pollution do comparable damage. For example, special legislation has been necessary to protect workers in certain industries, and this is generally so good that the greatest danger is probably to people living close by who are not so protected. Anyone interested in the nation's health will perform the most rewarding service by getting rid of the evils of smoking.

In almost all cases of outdoor air pollution, there are very good other reasons for getting rid of it, and the case on health grounds is not usually as compelling.

One of the most common difficulties is that in public and private discussion, pollution is attributed to the wrong source or an effect is incorrectly attributed to the pollutant. For example, brickworks at a distance of ten miles have been blamed for acid smuts such as are sometimes produced from the stacks of oil-fired boilers. These smuts had burnt holes in paintwork and fabrics. If they had come from the distant brickworks, their concentration at smaller distances would have been so great as to make life intolerable there. The smuts were observed at the same time as the well-known brickworks smells because they came from the same direction, and if this kind of evidence is adduced, a bad case is made, and a well-refuted, ill-founded complaint serves only to strengthen the position of the polluter.

Complaints of 'sulphur fumes' are often misguided. A public campaign was conducted against the power station whose chimney plumes are shown in Plate 5.6 largely on account of the whiteness of the plume.

It was incorrectly held responsible for much pollution from other sources because it could always be seen to be emitting at the time when fumes were detected. The City of Liverpool at that time contained many most disgraceful sources of pollution, the worst of which were domestic chimneys and ships. There was a permanent smelly haze of domestic smoke over some of the older housing areas, which reddened the sun and dirtied every object to the touch; yet the propagandists complained that the occasional shadow of the plume was cutting off much needed sunlight.

The total emission of sulphur was quoted to show that the power station was a major source of pollution in the city. Statements of total emissions are to be found in many documents, official and propagandist, as if the information could be useful on its own. One could argue that knowledge of the total emissions is a necessary starting point in discussions about pollution but this is absurd unless we have a certain means of deducing, from this knowledge, what concentrations of pollution will occur at ground level or in other places where it is important. In fact, we do not have such means, and our theories always have constants in them which have to be deduced by comparison with observations. If the theories were universally true so that from one set of observations, we could evaluate the constants and deduce what would happen on other occasions, the data on total emissions could be used, but this is not the case. The topography and the weather are so variable that we really have to study each case on its own, and if measurements cannot be made at the place of interest, a special study may be required in a very closely analogous situation. To provide the public with data of total emissions left in raw state, as part of a discussion of a serious problem, is damaging to progress. It may even hide some small, but insidious and damaging source in a column labelled "Others".

The question whether sulphur emissions on their own are damaging to health is an open one, but it seems probable that they are not. The whiff from the night-watchman's brazier is so much more intense than any other whiff of SO_2 received by the public that if its effects were harmful, the braziers would have been long ago condemned.

Traffic exhaust is objectionable. Because of a longer experience of petrol engine fumes, the public finds diesel fumes more unpleasant. Diesel engines need not produce black smoke, but as every cyclist knows, their exhaust is often very unpleasant. Nevertheless, the main argument against them may not be on grounds of damage to health. There is probably a serious effect due to carbon monoxide produced by petrol engines in streets with dense traffic or road tunnels. But CO cannot be smelled, and headaches and fractiousness it produces may be laid at other doors. An investigation in London bus garages showed that the health of employees in diesel bus garages was much better than in electric trolley-bus garages, and this was almost certainly because only in the latter was smoking permitted.

There is serious difficulty in using information about complaints either to condemn or to whitewash a source of pollution. In view of the complexity of the effects of dilution on smells already mentioned and other mechanisms of which observers may be ignorant, any authoritative suggestion tends to be taken as a general truth and this makes witnesses quite unreliable. Recently, I asked residents in a street close to an incinerator whether it was a cause of nasty odours, and people thereupon attributed to the public incinerator smells coming from a local electroplating works which was not mentioned in the conversation. Clearly, the wish to be helpful had some influence on their spontaneous replies, and it is much less interesting to have no information to impart.

The average citizen is too busy to bother to make official complaints about large industries, and does not wish to harm his neighbourly relations by complaining about smelly chimneys or bonfires. Recorded complaints are no measure of the magnitude of a nuisance, but more representative of a chance combination of the standards of the complainers and a large number of other factors such as vested interests, psychotic conditions, press campaigns, leading questions, and prejudices.

The outlook would be black, in view of all these and other uncertainties, if we did not really know what the serious sources of pollution were. The uncertainties enter the argument when pressures irrelevant to the facts of the pollution and its effects begin to be exerted. These pressures are, of course, not necessarily disreputable, because the wealth of the community depends very much on industry and transport which at present have to pollute the air. Progress is achieved by being aware of the mechanisms and processes by which pollution harms us so that we steadily eliminate the known worst sources and apply ever more stringent standards to new sources.

BLACKENING OF SURFACES

It is natural to regard the blackening of a solid surface exposed to the air as being due to the straightforward deposition of sticky substances such as domestic smoke on the surface. But more careful observation shows that many complicated mechanisms are involved, and some of these are scarcely understood.

Thermal precipitation (Plate 7.1) is a cause of much indoor blackening of walls and ceilings. There are two distinct mechanisms involved: first, the deposition occurs from warm air on to a cold surface. Consequently it is found above central heating radiators and hot pipes. It also occurs on ceilings, but often the deposition there is so uniform as to be imperceptible; however on some ceilings the non-uniform thermal conductivity of the ceiling is shown by the faint outlining of the rafters and hidden nail heads in the pattern of the deposition. It is prevented, or at least reduced by ensuring that the hot air is well mixed with room air before it makes contact with the wall.

PLATE 7.1. Thermal precipitation on to a cold wall and ceiling above a hot water pipe. The deposition occurs from the hottest air, and this is where the pipe is closest to the wall, or where some unevenness or obstruction to smooth flow stirs warm air from the convection current to the surface

PLATE 7.2. The north face of an octagonal tower on Exeter Cathedral is completely black while the adjacent faces remain relatively clean. The stonework is also blackened under eaves while the upward facing parts of the window stonework appear washed clean. Similar patchy blackening can be found on many buildings, old and new

The second mechanism which causes increased deposition is local turbulence in the airstream. This brings warm air more quickly to the surface and therefore increases the temperature gradient close to the surface. The turbulence also has the effect of carrying particles to the surface, and if they are sticky they remain on it. This last mechanism operates in the backs of our noses where the air we breathe passes through a narrow passage and emerges with considerable turbulence into a wider nazal passage in which dust particles are deposited on to a surface covered with mucous. The same kind of deposition due to turbulence in narrow air passages takes place around window and door frames, especially where they are not tight-fitting.

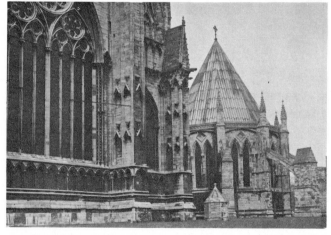

PLATE 7.3. The flying buttresses on Lincoln Chapter House show blackening in the 'shadows' of the columns at their outer ends. Also the south-east-facing wall is less blackened than the east- or south-facing ones, and this appears to be connected with the frequency of sunshine, the south-east face being much less shaded

These pictures were taken on an overcast day so that there were no sharp shadows

The effect of the airflow in determining the location of stone blackening is often clearly seen on the corners of buildings (Plate 7.2). There is probably not a single cause for this but almost certainly the separation of the airflow at corners plays a major part. The blackening is not usually a simple deposition, but is often organic and caused by the absorption of SO_2 from the air. There is a variety of chemical reactions and other mechanisms which induce or inhibit blackening of building stone or brick, among which are the following:

(i) the effect of wind in drying out the surface after rain.

(ii) the effect of wind in producing a differential wetting of the surface. Water may be an agent for cleaning the surface if it is sufficiently plentiful or for causing the blackening if it remains on the surface (Plate 7.4).

PLATE 7.4. Stone figures on Somerset House, London, where the upper surfaces are washed clean and the lower surfaces carry an accumulation or growth of black pollution. Similar disfigurements were common in the very soluble Headington stone used in Oxford before it was recently replaced

(iii) the possibility of frequently being very wet, thus intense blackening often occurs where water from an overflow pipe or leaky gutter often impinges on a wall, but where the stone is soluble it may be cleaner in such places.

(iv) the effect of shadows in delaying the drying of a surface by sunshine. Shadowed areas such as under window sills or north-facing walls often show more blackening (Plates 7.2, 7.3).

(v) the effect of direct sunshine in bleaching or preventing the growth of organisms.

(vi) the chemical nature of the surface, and in particular its solubility in a weak solution of the pollutant.

(vii) the physical nature of the stone surface, for example its power to hold water for long periods after wetting (Plate 7.6).

(viii) the presence of a (metallic?) catalyst for the chemical processes which cause blackening. Thus blackening often occurs

PLATE 7.5. Dust deposition on a figure in terracotta (non porous) on the Royal College of Science, London. This figure is protected from most rain by an overhanging balcony, and shows how darkening would occur under the effects of deposition alone. There is intense blackening in the places where no direct sunshine penetrates which may be due to a growth which is killed by sunshine

below an iron bolt head or nail in a wall, notably on red brick surfaces. This may be connected with the falling of rain down the wall at the protuberance, but appearances suggest a chemical effect.

(ix) the accumulation of black growth on the under side of stonework: this may be a mere deposition of material washed from the soluble parts, or a growth by chemical reaction in the presence of water which is favoured in shaded parts of the stonework.

(x) direct deposition of black particles undoubtedly takes place on to solid surfaces. But this can scarcely be the cause of blackening of vertical walls, and generally only occurs where the surface is not exposed to rain (Plate 7.5) which would wash off the deposition.

PLATE 7.6. Stonework on the south porch of Liverpool Cathedral which is built in red sandstone. The mortar between the stones seems to be sufficiently alkaline to prevent black growth: the stone surfaces vary greatly in some attribute, probably porosity, and have not been blackened uniformly. This stone is very susceptible to blackening where water runs down it, and this is detectable in this scene where rain running off the smooth mortar has caused a strip of blackening on the stone below

Often circular columns have one side blackened, and the edge of the blackening may be the line of separation of the airflow or the edge of the rain deposition when the wind is in the most common direction for rain. Sometimes it is the edge of the region not subject to sunshine. With a fuller knowledge of the causes of disfigurement by blackening it might be prevented. Various possibilities include the spraying of the stonework after washing with silicones or other non-wettable substances or with a solution of a chemical poisonous to the growth (this has obvious dangers).

EXPANSION AND CORROSION

Much corrosion is caused by natural pollution. Plate 7.7 shows one of the effects of airborne salt particles. Bulges and weakening are produced in overhead power cables which contain strands of aluminium and of steel. An electrolytic effect produces more rapid deterioration of the conductor cables than of the earth wires in general, but in coastal and badly polluted industrial areas the earth wires seem to suffer equally severely. The same is true of the stunting of trees and other vegetation—it is worse at the coast than in industrial areas, with a few localised exceptions.

The expansion due to reaction between a substance and an air pollutant often causes the flaking of stonework. Many cathedrals suffer from disfigurement of statues or decorated masonry because of this. Evidently the SO_2 in the air combines with carbonates of iron in some stones to form compounds, in the presence of water, of larger volume.

An interesting development in architecture is the decoration of large areas of wall of box-like buildings with some relief. A major reason for this is that streaks of blackening often cause unpleasant disfigurement of large uniform surfaces and the breaking up of the surface so that this disfigurement is not dominant, is attempted. Often, however, the decoration and relief are themselves disfigured in ways which are unpredictable because the mechanisms and reactions which cause disfigurement are scarcely understood.

PLATE 7.7. Examples of power cable corrosion from the coast of Cornwall. The upper picture shows a bulge due to the expansion of material in chemical reaction and the lower picture aluminium conducting stands taken from a point $1\frac{1}{4}$ miles from the sea where the effect was much less than close to coast. The four samples (left to right) show the outer and inner surfaces of an outer strand and the outer and inner surfaces of an inner strand

CHAPTER 8

Repercussions

THE phenomena so far described have been selected either
because they are fundamental mechanisms of atmospheric motion
and commonly occur, or because they have special importance
for air pollution, or because something fairly definite can be said
about them, or because they are very complicated and they have
up to now been treated rather naively. Thus thermal convection
is a basic atmospheric mechanism, fumigation is of major importance in altering ground level concentrations, the causes of and
mechanisms producing inversions and separation are fairly clearly
understood, and formulae which are correct in very restricted
circumstances only have been widely used as if they summed up
the whole spectrum of atmospheric behaviour.

From this we are given a picture of the atmosphere, governed
by the simple basic laws of physics and mechanics which are
understood and formulated with great precision, behaving in an
extremely complicated manner because of the great variety of
topography and climate. It is the complexity of the actual situation which creates the difficulties, not primarily the properties of
the materials which act out the weather. This is the feature of
meteorology which demands a much more mature approach than
the basic sciences. Above all we have to deal with the most
difficult branch of fluid mechanics—turbulence—made more
complicated than how engineers usually meet it by the all
important intervention of buoyancy forces.

The meteorology is complicated enough for anyone, and the
study of pollution in the air is made more complicated by the
disorderliness of the sources of pollution. The application of the
ideas and knowledge discussed in preceding chapters must involve

a detailed knowledge of the terrain and its weather. This is the context into which the pollution is born, and its behaviour and our reactions to it are absolutely determined by the context: the most ardent camper would not light a log fire on his drawing room floor.

There are NO universal formulae which can be used to describe what will happen to pollution emitted into the air, and this needs to be said because formulae have been so misused in the past by people who have been secretly frightened of realities and have dispensed according to the prescription of the most fashionable wizard. Responsibilities cannot so easily be shed. Pollution problems are certainly local, almost personal.

This book is not the place to go in more detail into the obvious recommendations not to make pollution at all, but if it must be made to render it as harmless as possible chemically. This of course is the only real solution to pollution problems. An ability to predict everything about the atmosphere would solve nothing because control of the atmosphere is out of the question. Any general recommendations a meteorologist can make must necessarily be extremely simple and really amount to this:

pollution should be emitted outside regions subject to stagnation under inversions, away from wakes of buildings, and at as great an altitude and with as much heat content as possible.

DOMESTIC PROBLEMS

The bonfire with its dense smoke and small production of heat is the most undesirable and unnecessary of all pollution sources. Ironically, about the only case for which a fairly reliable formula could be given is a bonfire in an infinitely large flat field on a very windy day; and this is no help at all! (Plate 8.1.)

But bonfires will be used occasionally because burning of refuse is often the easiest means of disposal and so the following recommendations are made:

(i) Bonfires should not be lit in the late afternoon or evening because the air is becoming stable. Because they are a ground

PLATE 8.1. 'Bonfire' smoke caused by the burning of grass on the embankment of an electrified railway line in a London suburb. This practice became customary when sparks from steam trains were liable to ignite dry grass growing uncontrolled, and produce dangerous conflagrations.

The objection to low level sources of smoke on windy days is that a large region is polluted before the smoke is dispersed upwards. Smoke producing material should be burned at a higher temperature with flames to burn the smoke, on a day of light wind so that the heat generated carries the gases out of the wakes of buildings

level source of pollution we must rely entirely on upward dispersal to keep the ground level air clean and so they should be confined to sunny hours around the middle of the day. (Plate 8.2.)

(ii) Bonfires should not be lit on very windy days because the smoke will be carried a large horizontal distance before it rises clear of buildings, etc.

(iii) Bonfires should not be allowed to smoulder: in the early stages they should be ignited with wood or other dry fuel which will burn with a hot flame, and all the refuse burned should be dry and should be placed on the fire in small quantities so as to avoid the production of smoke. *A good bonfire burns with tall flames.*

PLATE 8.2. Bonfire smoke trapped under an evening inversion. Because of the large temperature gradient in the evening and at night over cold ground the air entrained by a bonfire is much colder than that into which the smoke rises. Consequently even a good flaming fire on the ground cannot send its products very high and they are often trapped below tree and house tops

The effect is to generate the maximum amount of heat possible from the material being burned and to ensure that the smoke is burnt. Since smoke is usually unburnt fuel, a bonfire smokes because it is too cold and has not enough air. An improved draught usually eliminates smoke.

(iv) The amount of heat necessary to dry wet materials is very large because water has a large latent heat. The most economical way to dry things is in a wind, for in this way the heat needed to evaporate the water comes from the air. If they can be dried in the sunshine so much the better, but good ventilation is more important than heat in the drying process, as every experienced haymaker knows. Wet material to be burned should therefore be laid out close to the bonfire for drying before being burnt and not dried on the fire, or better still left to dry and burnt on another occasion.

(v) Bonfires should not be lit in the wake of a building, for the whole wake is thereby filled with smoke. Many gardens are so

PLATE 8.3. The smoke from a cottage chimney seen descending to the ground close to the house because it is not emitted so as to be outside the wake of the building. Because the chimney has no chimney pot the smoke is entrained into the chimney wake, and then into the wake of the roof. The only remedy is to have a tall slim chimney protruding two or three metres above the house top

enclosed that a bonfire should only be lit when the wind is light and then made hot enough for the gases to rise out of the wakes of houses.

(vi) Good gardeners know that garden refuse need never be burned. Twigs and other prunings which are not easily composted can always be left to dry before burning, and they then make a good basis for generating a hot fire in which potatoes may be roasted and fun had by all.

(vii) Bonfires are a very bad means of disposal of fabrics and of old mattresses in particular; the burning of rubber, paintwork, and plastics should be avoided because intense objectionable smells may be produced, and car batteries or other objects containing lead (e.g. good lead paint) produce poisonous fumes when burnt.

PLATE 8.4. A suburb of London drowned in its own low level emissions on an autumn afternoon. The air was so stable as a result of an anticyclonic inversion and the wind so light that the convection produced by sunshine only lifted the domestic smoke up to about 150 metres. The only solution in such a situation is to emit all necessary pollution at a height such that it would not be fumigated down to the ground

Domestic chimneys become less necessary with the wider use of electricity and of gas which produces only carbon dioxide and water vapour as combustion products. It is only a matter of time before fashion will relegate the attractions of the open fire as a means of heating, like the horse-drawn carriage as a means of transport, to the realm of Christmas cards. The chief domestic problem is the disposal of sulphur containing gases from heating systems using sulphur containing fuels, and most existing problems arise because architects are ignorant of them. They have regarded the chimney as a means of removing the flue gases from the interior of the house, and have ignored the surrounding air (Plate 8.3). From the aerodynamical point of view a chimney should be tall enough to emit its effluent, without downwash, clear of the wake of the building. Of course it is complained that such a chimney would be ugly, but this is a matter of opinion, and the complaint comes inappropriately from a community living in houses whose exteriors are cluttered with unsightly drainpipes and gas heater flues (Plate 8.5).

No determined effort to solve the problem of effective domestic chimneys has been made by architects or the building industry, and the tall stacks which may be seen on some Georgian and Victorian houses were built primarily to give good draught. The argument that high chimneys are ugly is utterly false: only the inducement to use them is lacking. Our houses are cluttered up with television aerials and advertisement hoardings when the need for them is imagined.

PLATE 8.5. Cottages for old people (modern version — Plate 8.3) of the nineteenth century with chimneys designed to produce a draught *and* to remove the smoke.

LARGE CHIMNEYS

All low level emissions are bad in principle (Plate 8.4). If a few large high level sources are to be substituted for a multitude

PLATE 8.6. Egborough power station in Yorkshire. The station stand in a wide plain with no pollution problems arising from the topography. The chimney is about 200 metres high and consists of four flues in one stack. By this means, when the station is on low load only one or two flues are used and the efflux velocity is sufficient to avoid downwash into the chimney wake. When on full load the heat content of the emission is concentrated and produces a greater thermal rise tl.an if the same effluent emerged from four widely separated chimneys.

It is possible that with smaller chimneys it is an advantage to separate them because in most wind directions their plumes would not overlap in the region on the ground where their maximum ground level concentrations are produced. But the separation has to be of the order of three or four chimney heights to obtain any real advantage in this way, and a separation of this magnitude in this case would be impracticable.

Modern chimneys are becoming so tall that the cloud-base inversion may tend to make any further increase in height unprofitable. Often, in moderate winds, the thermal rise will raise the effective height of this chimney to well above half way up to cloud base

of low ones the main possibility to guard against is the descent in moderate or strong, gusty winds. In that case, if the effluent is diffused along a cone, the source may be so large that the concentration may still be objectionably large when the plume meets the ground. Provided that this possibility can be prevented the

PLATE 8.7. Effluent from Kowloon power station (Hong Kong) accumulated under an inversion on a calm summer morning. Immediately above the station it rises higher, overshooting its equilibrium level and then falling back and spreading out. The station is conveniently placed at the water front so that fuel can easily be brought to it, but it is also close to the airfield so that its chimneys cannot be raised. In the polluted area the population density is extremely high and the pollution is a nuisance to the airfield. The only solution lies in the removal of the station to a greater distance where a very tall stack can be built

advantages of concentrating the emission are very great in all other kinds of weather.

It is possible to make rough calculations of the damage done by pollution of different kinds (see *J. Inst. Fuel* March 1957). If the total damage is known, the various features of a source can be used to assign factors which when multiplied together give their relative damaging power. Thus low level, low temperature sources in towns such as domestic fires and road traffic are at a disadvantage on every count when compared with isolated large hot sources (Plate 8.6). It is really immaterial whether the damage done by a ton of coal burned in the manner of Plate 8.3

is one thousand or ten million times as much as a ton used as in Plate 8.6: the factor is certainly very large: so large that in most countries if we suddenly changed from the worst to the best our pollution problem would be solved and a generation with new standards of air cleanliness would have to grow up before any problem would be recognised. Only problems due to mountains would remain.

Many unfortunate situations have arisen for economic reasons. Large industrial plants or power stations have been built on sites which happen to have been owned by the industry concerned and towns have grown up around them. A notable example is the power station at Kowloon (Plate 8.7) around which the population increased very rapidly in the period 1950-66 and the demand for electricity grew as a result. The airfield which is very near has also increased in importance as an international airport and it has consequently been necessary to restrict the height of the chimney. Power stations in big cities throughout the world are meeting similar problems, and very few are being built there for this reason. The development of very high power transmission lines makes it economical to build stations near to the source of fuel, which is the case at Egborough.

BACKGROUND POLLUTION

At distances of tens of miles when the air is not particularly stable the height of emission makes very little difference to the ground level concentrations produced. Background pollution, by which we mean that produced by distant sources, may become the main problem of future generations in densely industrialised countries, and the only solution is the limitation of emissions. (Plate 8.8). We began by reminding ourselves that the atmosphere is naturally polluted up to the tropopause, but that the cleaning mechanisms are so efficient that naturally produced pollution other than ordinary cloud is seldom a nuisance. The problems of civilisation will arise where the pollution is not soon washed out, and this will be mainly between latitudes 15° and 35° or in densely developed valleys among high mountains at higher latitudes. Many countries such as Britain, will need no solution better than the very tall chimney to satisfy most needs. But standards of cleanliness will rise and then events such as those seen in Plates 8.4 and 8.8 will not be tolerated.

PLATE 8.8. Pollution from Northern England trapped below an inversion seen at a distance of about 80 miles in the Isle of Man. The inversion on this occasion was at about 500 metres

INDEX

Numbers in bold refer to plates.

Adiabatic lapse rate 3
Agecroft power station **4.8**
Alps (French) **4.10–4.13**
Anabatic winds 76; **4.14**
Anticyclonic gloom 54; **4.5**
Architects 143
Area sources of pollution 40; **2.5**
Averages 20, 31

Background pollution 147
Bankside power station 86; **5.1**
Battersea power station 86; **5.2**
Bent-over plume 42
Bifurcation 43; **2.8, 5.2**
Black smoke **2.2, 2.10, 5.7, 5.8, 6.2**
Blackening of stone 129, 133; **7.2–7.6**
Blizzard 105; **5.18**
Blue haze 102
Blue plumes 95, 101; **5.14**
Bonfires 101, 139; **8.1, 8.2**
Brown plumes 101
Buoyancy in plume 41

Cape Cod **1.7**
Castellanus **1.17**
Cement works 91, 98; **2.1, 2.7, 4.6, 4.18, 5.5, 6.3**
Chamonix **4.15**
Chimney—
    idiotic 123; **6.9, 8.3**
    large 144; **8.6**
    pots 123
    wake **6.2, 8.3**

Cirrus 18; **1.18**
Cloud base (*see* Condensation level)
Cloud top 56; **3.1**
Cold inflow into chimneys 119; **6.7, 6.8**
Colour—
    of stratosphere and troposphere
    Frontispiece (*see* Blue plumes ; Red
    plumes ; *and* Plume, white)
Complaints about pollution 127, 129
Concentration—
    maximum 36
    of pollution in a plume 35
Condensation—
    in chimneys 90, 98
    of cloud 96, 101
Condensation level **1.11, 1.16**
Contrails 97; **5.10**
Convection over sea 56; **3.1**
Cooling from below 56
Cooling towers 91, 96; **1.12, 5.4, 5.9**
Corrosion 136; **7.7**

Dacca **1.4**
Descent of wet plumes 87; **5.1, 5.3**
Desert locusts 106; **5.20**
Dew 7
Diffusion in a plume 29
Diurnal temperature variations 3
Domestic smoke **1.3, 1.4, 2.5, 4.5, 5.14, 6.6, 8.3–8.5**
Dovey estuary **1.9**

Downwash—
    on chimneys   110; **2.10, 6.2, 8.3**
    on cooling towers   **5.4**
Drizzle   89
Droplets   88
    fallout of   89
    removal of   86
Dust—
    deposition   24; **7.5**
    devil   103; **5.16**
    gauge   25; Fig. 4
    haze   105; **5.17**

Eddies—
    causes of   22
    effect on plume width   31
Edinburgh   **2.5**
Egborough power station   **8.6**
Equilibrium level   41; **2.3, 4.9**
Evaporation—
    of cloud by convection   56; **3.2**
    of cloud by subsidence   57
    of moist plumes   86–92, 101; **2.1, 5.3**
Exhaust from vehicles   128

Fog—
    dispersal   10; **1.9**
    formation   98
    mountain   **4.1, 4.2**
    radiation   3–6, 18; **1.1, 1.2**
    sea   7; **1.6, 1.7, 1.9**
    valley   **4.3**
Fumigation   13, 19; **1.14**

Gas washing   86; **5.1, 5.2**
Ground, effect on concentration   34
Guttation   7; **1.8**

Haboob   105; **5.19**
Hams Hall power station   **1.12**
Handley-Page slot   117; Fig. 25
Harmattan   105; **5.17**

Health, effect of pollution on   126
Hillside pollution   80

Inverse square law   36
Inversion—
    cloud base   14, 64; **1.15**
    cloud top   56
    behind fronts   67; **3.8**
    high level   54
    nocturnal   **1.5**
    penetration of   58; **3.3**
    permanence of   63
    behind showers   67; **3.9**
    made visible by smoke   **1.15, 2.3, 3.4, 3.5, 8.2, 8.8**
    valley   70
Ironbridge power station   **2.4, 4.9, 5.12**
Isle of Man   40; **8.8**

Jura mountains   **4.16**

Karachi   **1.5, 5.17**
Katabatic winds   70; **4.2**
Kent   **1.15**
Kowloon   147; **8.7**

Lapse rate—
    dry adiabatic   3
    wet adiabatic   14
Lebanon   **4.18**
Lee eddy   112
Lime kiln   **2.10**
Liverpool   127
Los Angeles   83; **4.20**

Milford Haven   **1.14**
Mirage   7; **1.10**
Mount Isa   **2.3**
Much Wenlock   **1.3**

Natural pollution   102

Neutral stratification 3
New Zealand 83; **4.17, 4.19**
Norwich **1.1**

Oil refinery **1.14**
Open fires 143

Plume—
  bent-over 42
  near buildings 52
  dark **5.2, 5.5**
  disrupted by eddies **2.4, 5.7**
  at inversion **2.3, 4.9**
  effect of inversion 39
  narrow **2.1**
  sinuous **2.2, 2.7**
  spread 29
  wet 86; **5.1–5.3**
  white 86; **5.6**
Power stations **1.1, 1.12, 2.4, 2.6,
  2.8, 3.3, 4.8, 4.9, 4.17, 5.1, 5.2,
  5.4, 5.9, 5.10, 5.12, 8.6, 8.7**

Radiation—
  from clouds 18, 56
  from fog 70
  from pollution 64
  from subsiding air 55
Red plumes 99; **5.13**

Salient edge 108, 111; **6.5**
Saltburn **5.15**
Salt haze 102; **5.15**
Sampling time 38
Santiago, Chile **4.21, 4.22**
Separation 107; **6.1, 6.4, 6.5**
  edge of blackening 133, 136; **7.2**
  on buildings 114
  on cliffs and mountains 111; **6.1,
  6.4**

Smells, dilution of 124
Smog 19, 81; **3.3, 4.7, 4.8, 4.19–4.21**
Smoke definition 99
Smoke layers **1.5, 1.14**
Smoke trails undispersed **1.3, 1.4,
  2.3**
Snow line 76; **4.12, 4.13**
Stability 1
Stable air 3
Steaming water 97; **5.11, 5.12**
Steam trains **5.3**
Steel works 99; **2.9, 5.13**
Stratification of air 2
Stratosphere *Frontispiece*
Subsidence 54
Sulphuric acid plumes 93; **5.6**
Sulphur dioxide 127, 128, 136

Thermal convection 1, 7; **1.13**
  artificial 10; **1.11, 1.12, 2.9, 5.9**
  effect on pollution 21
  sources 51
Thermal precipitation 130; **7.1**
Thermal rise of a plume 46
Total emissions 127
Tropopause *Frontispiece*

Unstable air 7

Valley fog **4.3**
Valley pollution 70; **4.10–4.13**

Wake—
  of a body (*see* Separation) 108
  of a building 115, 118
  of a hill 112; **6.3**
Wave cloud **3.4**
Welsh coast **1.6, 1.9, 3.4**
White smoke 101; **5.6, 6.6**
Wind speed, effect of 21